REPORT

BY

DR. M. C. COOKE

ON THE

GUMS, RESINS, OLEO-RESINS, AND RESINOUS PRODUCTS IN THE INDIA MUSEUM

OR

PRODUCED IN INDIA

REPORT

BY

DR. M. C. COOKE,

ON THE

GUMS, RESINS, OLEO-RESINS, AND RESINOUS PRODUCTS

IN THE INDIA MUSEUM

OR

PRODUCED IN INDIA:

PREPARED UNDER THE DIRECTION OF THE REPORTER ON THE PRODUCTS OF INDIA

M. C. COOKE

GYAN

Published by

GYAN

Gyan Publishing House
5, Ansari Road
Daryaganj, New Delhi-110002
Phone: 011-47034999, 9811692060
E-mail: books@gyanbooks.com

Distribution Network
gyanbooks.com
India, USA, Canada, UK, Australia

ISBN: 978-93-6433-303-0 (HB)

First Published, 1874

2nd Impression 2025

Printed at: Gyan Press, Delhi.

ON THE GUMS, RESINS, OLEO-RESINS, AND RESINOUS PRODUCTS IN THE INDIA MUSEUM OR PRODUCED IN INDIA

Author: M. C. COOKE

REPORT

BY

DR. M. C. COOKE,

ON THE

GUMS, RESINS, OLEO-RESINS, AND RESINOUS PRODUCTS

IN THE INDIA MUSEUM,

OR

PRODUCED IN INDIA.

PREPARED UNDER THE DIRECTION OF THE REPORTER ON THE PRODUCTS OF INDIA.

LONDON:
INDIA MUSEUM.

1874.

INDEX.

INDIAN GUMS AND RESINS.

In submitting Report on the Gums and Resins of India, it will be necessary first to explain the classification employed. Hitherto no classification has been attempted of these products, and in all lists prepared a merely mechanical or alphabetical arrangement has been adopted. The method which commended itself as most serviceable for commercial purposes was one which bore some relation to the function of the substances under report, a method already adopted in respect of East Indian fibres, and which practically has been found to answer its purpose. In following this plan, the first grouping that suggested itself was derived from the principal characters of the substances themselves, features which many would have in common, and which would in no way interfere with the integrity of the subordinate groups. These four principal groups are Gums, Gum Resins, Resins, and Oleo-Resins. Their names will at once convey the limits of each of these primary divisions.

The subsidiary groups, with one example enumerated as a type of each, will be seen at a glance in the following Table :—

I.—Gums.

- A. True Gums.
  - α Arabic kind [Acacia Arabica].
  - β Cherry kind [Prunus cerasus].
- B. Pseudo-Gums.
  - α Tragacanth kind [Stercula urens].
  - β Dark or Moringa [Moringa pterygosperma].
- C. Astringent Gums [Butea frondosa].

II.—Gum Resins.

- A. Emulsive [Gamboge].
- B. Fœtid [Asafœtida].
- C. Fragrant.
  - α Bdellium kind [Googul].
  - β Benzoin kind [Benzoin].

III.—Resins.

- A. Hard or Copalline.
  - α Pale resins [Vateria Indica].
  - β Dark resins [Black Dammar].
- B. Soft or Elemi [Canarium commune].

IV.—Oleo-Resins.

- A. Balsams [Wood oil].
- B. Varnishes [Burmese Lacquer].
- C. Turpentine and Tar.

These groups require very little explanation beyond that which the table affords. In our present imperfect acquaintance with some of the true gums it has been found necessary to place them alphabetically under this group, inasmuch as their separation into the two kinds of Arabic gum and cherry gum could not be wholly completed in default of specimens for examination. It may be remarked that the difference contemplated in this division lies in their comparative solubility in water, those of the Arabic kind being wholly soluble, whilst those of the cherry kind swell considerably in water, but only partially dissolve. The latter are sometimes called Bassorin gums.

Of the pseudo-gums which swell in water but do not dissolve, one kind partakes of much of the character of Tragacanth in being almost colourless, the other dark coloured, and as far as at present known commercially valueless.

The third group is introduced here provisionally, but it is doubtful whether such substances as kino, either that derived from *Pterocarpus marsupium* and its allies, or from *Butea frondosa* and its allies, should not be relegated into the company of such extracts as Cutch and Gambier, although commonly classed with gums.

Of gum resins, the emulsive series including gamboges of all varieties will follow the gums in natural order in consequence of the facility with which they mix and form an emulsion with water.

The Fœtid gum resins is a very natural and complete group, since all are medicinal and the majority are derived from Umbelliferous herbaceous plants. There is also a great affinity of chemical characteristics.

The Fragrant gum resins hardly constitute such a satisfactory group, which necessarily has been subdivided into two sections, the one to include substances of the Myrrh and Bdellium kind, more or less bitter to the taste, and the other the truly fragrant gum-resins, such as gum Benzoin, Olibanum, and other incense gums.

Of the true resins, the first group contains the comparatively hard resins resembling Copal, applicable to similar purposes. The commercial term of *Hard Copal* is much more restrictive, and its use is not contemplated here. Hard resins might be subdivided again into hard and soft copals. None of the Indian resins belong to the commercial "Hard Copal." The division into Pale and Dark resins will again represent comparative value, since the Black resins would not be available for the same purposes as the Pale resins.

Elemi resins have a different chemical composition, and distinct uses, mostly medicinal, and are not available for varnish manufacture.

Oleo-resins consist mainly of a resin suspended in volatile oil, and the three groups here proposed are in reality almost functional, for the Balsams have little value save in medicine. Natural varnishes for coating wood and iron have little other use, and turpentine with tar and rosin complete the series.

The materials upon which this report are based consist of a small reference collection, derived from various sources, not confined to India, and in most instances well authenticated, which is relied upon chiefly as an aid to identification. Manifestly it is impossible in a great many cases to determine the source of a gum or resin by mere comparison of specimens, but authenticated specimens are never wholly useless as collateral evidence.

Unfortunately the gums and resins which were exhibited at the Universal Exhibition of 1851 by the late Dr. Royle are deficient from the collection, the "Drugs" of which are now in this department.

The second source of material is the Museum Collection, which includes specimens sent from India to the Exhibition of 1862 and all European Exhibitions since that period. Unfortunately a great number of specimens have always been sent without intimation of their botanic sources, and only with vague, sometimes illegible, native names. Hence, this collection is by no means so complete as it might have been. Recently it has been augmented from the samples sent over from time to time for valuation and report.

As to the information embodied in this report, and details of collection, &c., where given, the authorities are in all cases cited, so that it is unnecessary to recapitulate them here. The statistics are derived from the trade returns.

Group I.—GUMS.

Before entering upon the examination of the numerous kinds contained in this group, it may be premised that there are certain elements essential to a commercially good gum, as well as to any other trade product, and that it is unreasonable to suppose that gums which do not fulfil these conditions can compete in the markets with those which do. One of the first conditions is the entire solubility of the gum in cold water, so as to form a smooth and homogeneous mucilage. Gum which does not wholly dissolve in cold water, but which would dissolve by the application of heat, though not of the first class, would still commend itself, if other conditions were fulfilled. There are purposes to which cherry-gums and the Tragacanths are applied, but these are much less numerous or important in their demands than for a good "Arabic gum."

2. A good gum should be pale in colour. It may be true that a good and perfectly soluble gum, like "Babool," for instance, may form as smooth and tenacious a mucilage as the best Turkey Arabic, but, because dark coloured, and for no other reason, would have a much lower marketable value.

3. It is scarcely necessary to add that the gum must be unmixed. Two gums, it even may be that both are soluble, might not dissolve equally in rate, or freely coalesce; this would be a disadvantage, but when two or more gums are mixed, of which one is insoluble in whole or in part, the whole sample is greatly deteriorated,

and reduced to almost the rate of the insoluble admixture. Hence then, commercially, it is of importance that gum should be (1) entirely derived from the same species of tree, (2) as light and uniform in colour as possible, and (3) free from all foreign admixture. There are two kinds of admixture of foreign matters to which gums are liable, one accidental, and the other fraudulent. The accidental include bark, chips, &c., to which the gum is adherent when collected, and often to be attributed to careless collection. The fraudulent consists in the addition to what would otherwise be a good gum of foreign substances in order to increase the weight. We have seen a good pale Arabic, though small, mixed with such a large proportion of sand in order to increase its weight as to render it absolutely unsaleable, except at a nominal rate. The gum was broken down to cover the deception, and hence the sand could not be sifted from the gum except by sacrifice of more than an equal bulk of powdered gum.

These facts must be borne in mind in all attempts to estimate the market value of gums. The highest type is undoubtedly "picked Turkey Arabic," but there are many good "Arabic" gums which do not reach this standard. It is always well to keep in view the commercial type of the highest and best quality of any product, in order to measure against it any other competing substance, and thus estimate the amount and direction of divergence from the given type.

Sect. A.—TRUE GUMS.

Acacia vera. *Willd.*

Willd. Sp. Pl. iv. 1084. *DCand. Prod.* ii. 461.
Mimosa nilotica. *Linn. Sp. Pl.* 1506.

A middling sized tree, with spines in pairs, straight, sharp, about a quarter to half an inch long. Leaves bipinnate, and as well as the branches smooth, these are covered with a reddish brown bark. Two pairs of pinnæ, leaflets small, 8 10 pairs, oblong linear, with a gland between the pinnæ. Flowers in yellow globose heads from two to five in axillæ of the leaves and stalked. The legume is moniliform, short, straight, and containing but few seeds.

Hab.—Egypt to Senegal.

Samaghe-aarabbi, *Ar.*

This tree yields the usual Arabic gum of commerce, which finds its way into India by Bombay from Aden, and other places on the Arabian coast. Vaughan says "it is "known in Arabia and on the African coast by the name of Samgh. The acacia "which yields this gum is generally a small shrub of a dry and withered appearance; "occasionally, however, it shoots out into a tree of from 20 to 30 feet high. The "Somalis on the north-east coast of Africa collect the gum during the months of "December and January. The process of obtaining it is extremely simple, long "incisions are made in the stem and branches, from which the juice flows, and when "dry is removed. After the gum of a district has been gathered, it is sewn up in "goat skins, and brought on camels to the great Berbera fair, or to some of the small "settlements on the coast, and thence shipped for Aden and India. Three descriptions "of the gum, styled severally Felick, Zeila, and Berbera, are exported from the Somali "coast.

"*Felick gum* is collected chiefly by the Magartam Somalis and those who inhabit "the district of Gardaf or Cape Gardufui. None of this quality, which is esteemed "the best, finds its way to Aden, a little reaches Maculla and Shehr on the Arabian "coast, but the mass is usually bought up by the Banians and shipped direct for "India. The gum felick realises about 25 rupees the hundredweight in the Bombay "market.

"*Zeila gum*, so called from the port of that name, and *Berbera gum*, from the "district of Berbera. The former of these is sold for 15, and the latter for 13 "rupees the hundredweight in the Bombay bazaar.

"The Acacia is common through Yaman and Hadramant, but the Arabs do not "appear to appreciate the shrub, as but very little gum is collected by them. "The gum forms, however, an important ingredient in the preparation of their "writing ink, to which it gives that glossy appearance so much admired in old "Arabian manuscripts. They also use it occasionally as a demulcent and nutrient, "and give it in the form of mucilage to invalids as we do arrowroot, sago, &c. The "Somalis resident at Aden occasionally go a short distance into the interior and "gather small quantities, which they immediately sell in the bazaar. The natives on "the south-east coast of Arabia, between Aden and Maculla, also collect a little, but "of this scarcely any is exported. During 1851, 250 tons of gum arabic passed

" through the Aden custom-house. The selling price here is from 2¾ to 3 rupees the " maund of twenty-eight pounds." (*Pharm. Journ.* xii. (1852) 226.)

Flückiger and Hanbury state that the gum arabic of commerce is furnished by *Acacia Verek.*

The two samples in the Museum Collection are (No. 4030) from South Kensington and (5200) from Vienna Exhibitions of 1873, derived from Aden *viâ* Bombay, No. 1128, called Babool Gum, from Aden, is probably a coarser darker sample of the same gum.

IMPORTS OF GUM ARABIC INTO BOMBAY PRESIDENCY.

——	1869–70.	1870–1.	1871–2.	1872–3.
From Suez	32	- - -	1	—
" Turkey	- - -	- - -	- - -	18
" Africa, Coast	- - -	- - -	- - -	136
" Red Sea	4,855	8,146	8,083	13,106
" Aden	4,814	1,576	1,619	927
" Persian Gulf	98	6	157	165
Total cwts.	9,799	9,728	9,860	14,352
From Bengal Presidency	- - -	3	—	—

EXPORTS OF GUM ARABIC FROM BOMBAY PRESIDENCY.

——	1869–70.	1870–1.	1871–2.	1872–3.
To United Kingdom	- - -	4,347	2,126	4,561
" France	- - -	- - -	- - -	60
" Genoa	- - -	- - -	26	—
" Trieste	- - -	- - -	58	3
" Africa, Coast	- - -	5	- - -	1
" North America	- - -	126	60	—
" China	- - -	93	35	—
" Japan	- - -	- - -	120	—
Total cwts.	- - -	4,571	2,425	4,625
To Madras Presidency	- - -	1	5	—
" Bengal Presidency	- - -	287	—	—

Acacia Arabica. *Willd.*

Willd. sp. Pl. iv. 1084. *DCand. Prod.* ii. 461.

Mimosa Arabica, *Lam.*; *Roxb. Fl. Ind.* ii. 557; *Cor. Pl.* ii. 149.

Tree 30–40 feet, armed with stipulary thorns, leaves bipinnated, pinnæ about five pair, leaflets 15–20 pair, glabrous, peduncles aggregated, axillary or forming a raceme by the abortion of the leaves; heads of flowers globose; corolla 5 cleft; stamens distinct; legumes stalked, thickish, contracted on both sutures between the seeds; flowers small, bright yellow, fragrant.

Hab.—Bengal. Coromandel. Deccan.

The native synonyms for the gum are thus given by Moodeen Shereef:—

Samaghraarabbi. *A. P.*
Babul-ki-gond. *H.*
Kikar-ka-gond. *Duk.*
Velam-pisin, Karu-velam-pisin. *Tam.*
Nalla-tumma-banka, Tumma-banka. *Tel.*
Velakam-pasha, Karu-velakam-pasha. *Mal.*
Gobbali-gondu, Kare-gobbali-gondu. *Can.*
Babuler-gun. *Beng.*
Kula-barbura-niriyasam. *Sans.*
Kula-babli-cha-gonda. *Mah.*
Kulo-baval-nu-gundar. *Guz.*

Babool gum is usually in rather large tears or portion of tears of a more or less dark brown colour, often mixed with impurities, rather brittle, of a shining fracture wholly soluble in water, forming a dark coloured mucilage.

In the Museum Collection the best specimen is from Bombay, of an amber colour in rounded tears, evidently a picked sample. There are also good specimens from Rajpootana, a darker sample from Bombay, a good pale gum from Mysore. Dark coloured, but strong gums from Ahmedabad, Gyah, Bengal. Two rather peculiar specimens in small tears from Jhansee and Lucknow. Two samples, also in small tears, from Calcutta and Ahmednugger, are mixed gums, containing a large proportion of what appears to be the gum of *Conocarpus latifolius*. A dark bright gum (not identified) from Chota Nagpore seems to be a Babool gum, as also a specimen from British Burmah. "Royana" gum (4025) from Baroda, resembles strongly an Acacia gum, but whether it is this species cannot be determined.

Acacia catechu. *L.*

DCand. Prod. ii. 458; *Wi. & Arn. Prod.* i. 272.
Acacia polyacantha, *DCand. Prod.* ii. 459.
Acacia Wallichiana, *DCand. Prod.* ii. 485.
Mimosa catechu, *Roxb. Fl. Ind.* ii. 563.
Mimosa catechuoides, *Roxb. Fl. Ind.* ii. 562.

Tree. Branches armed with stipulary thorns or occasionally unarmed, pinnæ 10–30 pair, leaflets 30–50 pair, petiole sometimes armed on the under side with a row of prickles with one large gland below the lowest pair of pinnæ and between the extreme 1–6 pairs; spikes axillary, 1–4 together shorter than the leaves; petals united; stamens numerous, legumes flat, thin, straight, linear, glabrous, 4–8 seeded; flowers numerous, small, white.

Hab.—Coromandel, N. Concan.

The native synonyms for the tree are:—

Kheir, Khaira, K'hayar. *H.*
Wothalay. *Tam.*
Khuera. *B.*
Khehiree. *Cing.*
Podeel-maum. *Tel.*
Khadera-kair. *Mahr.*
Shabin. *Burm.*

Kheir gum is in rounded tears varying from the size of a pea to that of a small walnut, or in broken fragments, mostly of bright coloured shades of dark amber or mahogany brown, rather friable, the tears being cracked and granular resembling coarse brown sugar. Sweet to the taste, especially the darker tears, quickly soluble in water, yielding a thin but strong mucilage of a deep brown sherry colour.

Sample of this gum was sent over from Chanda in 1873 for valuation and report. The Deputy Commissioner then remarked that mixed gum is sold in the bazaar at Chanda at about Rs. 4 for the maund of 80 English pounds or Rs. 5–8 the cwt. The cost of collecting the gum is estimated at Rs. 2¼ the maund, carriage to the railway 8 annas, and carriage by rail to Bombay 12 annas. The cost price of Chanda gum may therefore be put down at Rs. 3 · 12 the maund of 80 lbs., or say Rs. 5 · 4 per cwt. at Bombay.

The one disadvantage of sending very pure specimens is that it is frequently doubted whether the article can be had in large quantities similarly pure. Of course if merchants who desire to purchase these gums expect to find them in the bazaar in this condition, they will be disappointed. No gums can be purchased in the bazaar here pure and unmixed with other gums, but if merchants would take some trouble in the matter they would soon be able to obtain this gum quite as pure as the sample and in large quantities.

The valuation of the gum sent was reported "ordinary Arabic value 20*s.* to 25*s.* "per cwt."

The specimens of this gum in the Museum Collection are from Goonah, Indore, Jhallawar, Ahmedabad, Chanda, Nagode, and Chutterpore, the two latter inferior.

Acacia speciosa. *Willd.*

DC. Prod. ii. 467. *Wi. & Arn. Prod.* 275.
Mimosa speciosa, *Jacq. Coll.* i. 47.
Mimosa sirissa, *Roxb. Fl. Ind.* ii. 554.
Mimosa flexuosa, *Rottl.*
Acacia Lebbek, *Willd. DC. Prod.* ii. 466.
Albizzia Lebbek, *Benth. Hook. Journ.* iii. 87.
Albizzia latifolia, *Boivin. Ency.* xix. ii. 33.

Unarmed tree; pinnæ 1–4 pair, with a large gland about the base of the petiole; leaflets 4–9 pair, oval, unequal, often with one or two small glands near the base of the partial petioles and small ones between the leaflets; peduncles axillary, 1–4 together, each bearing a globular head of shortly pedicelled flowers; calyx long, tubular, petals united to beyond the calyx; stamens very long, monadelphous; legume thin, flat, broad, linear, remotely 8–10 seeded; flowers small, white, fragrant. (*Drury.*)

The native names of this tree are—

Sirissa, Siris. *B.*	Sirus. *Dec.*
Dirasana. *Tel.*	Siris. *H.*
Vel-venge-katuvagi. *Tam.*	

Drury observes that this tree yields a considerable quantity of gum, which is valuable for many ordinary purposes. It is something like gum Arabic. Dr. Stewart writes that the gum is similar in quality to gum Arabic, but I do not know of its being collected or sold in quantity. (*Stew. Punj. Pl.* 56.) A sample from Kurnool (No. 1727) can scarcely be distinguished from a good Babool gum.

In the Pharmacopœia of India the gum is said to be of a superior quality. Whereas Mr. Baden Powell in "Punjab Products" remarks that "Siris" gum is not soluble entirely but softens into a jelly-like size. There is a specimen in the Museum Collection (No. 2071), which accords with this latter description; it is of the cherry gum kind, in dull irregular tears of a dark brown colour and tasteless, swelling in water, but only partially soluble. It is referred to *Acacia lebbek*. It is further remarked in "Punjab Products" that the Siris gum is "a coarse gum used for adulterating gum-"Arabic and used under the name of 'lera' in the art of printing calicoes with gold "and silver leaf patterns. It is not soluble completely, like 'gond Arabi,' but forms "a stiff kind of jelly."

The Museum samples are from Kurnool, Salem, and Madura, the latter of very inferior quality.

Acacia modesta. *Wall.*

Wall. Plant As. Rar. ii. t. 130.

Shrub. Spines axillary, subulate; pinnæ 2–3 pair; leaflets 4 pair, oblong, obtuse, glaucous; petioles pilose, with one gland about the middle. Spikes axillary and terminal, usually solitary, nodding on short peduncles; legume stalked, linear-oblong, flat, 6 seeded, acute at both ends; flowers white, fragrant.

Hab. In dry tracts between Saharunpore and Delhi.

The native synonyms of this gum are—

Bhimbri-gond. *H.*
Gond-phulahi. *Punj.*

Dr. Stewart remarks that this tree yields sparingly a gum similar to gum Arabic, which Bellew states the people of the Peshawur valley consider to be restorative. Whereas Powell adds "it is in little curled pieces quite soluble and of a yellow colour." *Punj. Prod.* p. 415.

Acacia odoratissima. *Willd.*

DCand. Prod. ii. 466.

Albizzia odoratissima, *Benth. in Hook. Jour.* iii. 88.

Albizzia micrantha, *Boivin. Encycl.* 2 p. 33.

Mimosa odoratissima, *Roxb. Fl. Ind.* ii. 546; *Cor. Pl.* t. 120.

Acacia lomatocarpa, *DC. Prod.* ii. 467.

Large tree. Pinnæ 3–8 pair, leaflets 8–25 pair, broadly oblong, very unequal sided, glaucous beneath; panicle many headed; flowers pubescent, pale yellowish, fragrant, legume flat, broadly linear, thin, margins thickish, about 10 seeded. (*Drury.*)

Hab.—Coromandel. Concans. Neilgherries. Assam.

Native synonyms for the tree are—

Curroovaga, Karuvenge. *Tam.*	Karinthakara. *Mal.*
Shinduga. *Tel.*	Ran-sirrus. *Dec.*

A specimen of gum from Salem, referred to this species in shining rounded tears, liable to agglutinate, of a dark brown colour, resembling Babool gum, tasteless and soluble, is the only specimen in the Museum Collection (Vienna, 1873).

Acacia ferruginea. *DC.*

DCand. Prod. ii. 458. *W. & Arn. Prod.* 1. 273.

Mimosa ferruginea, *Roxb. Fl. Ind.* ii. 561.

Tree armed with conical stipulary thorns, occasionally unarmed; pinnæ 3–6 pair with one gland on the petiole and one between each of the 1–2 extreme pairs; leaflets 10–20 pair, oblong linear, obtuse; spikes axillary, usually in pairs, stamens slightly united at the base; legumes rusty coloured, flat, lanceolate, hard, 2–6 seeded; flowers small, pale yellow.

Hab.—Circars. Courtallum.
Native synonyms for the tree—

Vel vaila, Velvaylum. *Tam.*
Woanee, Tella-toomma. *Tel.*

The sample of gum in the collection from Madura (2052) is in dark brown shining fragments of large tears, moist, soon agglutinated and tenacious, soft, tasteless, soluble in water, forming a coloured mucilage. A specimen sent to the Madras Exhibition of 1855 by Dr. Cleghorn is characterised as an inferior gum.

Acacia leucophlæa. *Willd.*

Wi. & Arn. Prod. l. 277. *DCand. Prod.* ii. 462.
Mimosa leucophlæa, *Roxb. Fl. Ind.* ii. 558; *Cor. Pl.* t. 150.

Tree armed with stipulary thorns; pinnæ 7–12 pair, with a gland below the first and between some of the last pairs; leaflets numerous, oblong-linear, pubescent or nearly glabrous; panicles large, terminal, or from the upper axils; branches and peduncles shortly tomentose; heads of flowers globose, on short peduncles, legume narrow, linear, long, curved, shortly tomentose when young; flowers pale yellowish. (*Drury.*)

Hab.—Coromandel. South Mahratta country. Sholapore. Delhi.
Native synonyms for the tree—

Kikar. *H.*
Jumma-jegota. *B.*
Vel-vaylum. *Tam*
Tella-toomma. *Tel.*

In Ure's dictionary it is stated that "Gum Bassora is a gum obtained from *Acacia* " *leucophlæa* brought from Bassora. It has a spec. grav. of 1·3591, and is yellowish " white in colour."

Acacia sundra. *DC.*

DCand. Prod. ii. 458. *Wi. & Arn. Prod.* i. 273.
Mimosa sundra, *Roxb. Fl. Ind.* ii. 502. *Cor. Pl.* t. 225.

Tree, everywhere glabrous; branches armed with decurrent recurved stipulary prickles, sometimes entirely unarmed; pinnæ 15–20 pair, with a gland on the petiole below the lowest pair and between the extreme pairs, leaflets 20–40 pair, small, linear obtuse; spikes 1–3 together, axillary, shorter than the leaves; stamens distinct, legumes flat, lanceolate, few seeded, occasionally emarginate between the seeds; flowers small, yellow.

Hab.—Mountains of Coromandel. Soonderbunds.
Synonyms for the tree:—

Sandra, Chandra, Nella-sandra Tella-sundra. *Tel.*
Karangalli, Baga, Caroongally. *Tam.*
Lall-kheir. *H. & Mahr.*
Khoira. *Uriya.*

Whether the gum yielded by this tree is to be obtained in any quantity we have no evidence to show. The specimen in the reference collection more closely resembles a dark Babool than the Kheir gum, though the tree itself is closely allied to *A. catechu.*

Adenanthera pavonina. *L.*

Wi. & Arn. Prod. i. 271. *DCand. Prod.* ii. 446. *Roxb. Fl. Ind.* iii. 370. *Wi. Ill.* i. t. 80.

Tree unarmed; leaves bipinnated; leaflets oval, obtuse, glabrous on both sides; racemes spike-like, terminal, or from the axils of the upper leaves; legume somewhat falcate and twisted; flowers small, yellow, fragrant. (*Drury.*)

Hab.—Peninsula. Travancore. Sylhet. Assam.
The native names of the tree are:—

Ranjuna, Rukta-chundana Rukta-chundun, Ranguna. *B.*
Koochunduna. *H.*
Aunay-coondoomany. *Tam.*
Bandi-gooroovindza Manseni-kotta. *Tel.*
Manjatie. *Mal.*
Mandateea. *Can.*
Mansiadi. *Cyn.*

This tree yields in Ceylon a gum called "Madatia," but whether such gum is produced or known in India there is no evidence to show.

Ægle Marmelos. *Corr.*

Wi. & Arn. Prod. i. 96; *DCand. Prod.* 1. 538; *Roxb. Fl. Ind.* ii. 579; *Wi. Icon.* t. 16; *Roxb. Cor. Pl.* ii. 143.

Feronia pellucida, *DC.* Prod. i. 538.

Cratæva marmelos, *Lam.*

Tree with pinnate leaves; leaflets three, seldom five, middle one petiolate, lateral ones almost sessile, oblong or broad lanceolate, crenulated, the terminal one the largest peduncles axillary, few flowered; pedicels long, seeds covered with a transparent glutinous matter; flowers large, greenish-white. (*Drury.*)

Hab.—Malabar and Coromandel.

Native synonyms:—

Bel, Bael. *H.*
Vilva, Velva. *Tam.*
Bel. *B.*
Bel. *Mahr.*
Maredoo, Vilva-chettu, Bilvamu-chettu, Maredu, Maluramu-chettu. *Tel.*
Kuvalam, *Mal.*
Belloo. *Uriya.*
Beli. *Cing.*
Ouk-sheet. *Burm.*

Amongst the gums contributed by Dr. Cleghorn to the Madras Exhibition of 1855 was the gum of this tree, which was styled a "good Arabic." The sample in the Museum Collection is from Shemooga. It is an amber coloured gum, in small fragments of tears, resembling in colour and appearance a coarse raw sugar.

Amygdalus Persica. *L.*

Roxb. Fl. Ind. ii. 500.

Tree: branches spreading, leaves lanceolate, glabrous, serrated; flowers sessile with reddish calyces and pale or dark red corollas. Drupe roundish, generally pointed, with a longitudinal groove; seed hard, irregularly furrowed.

Hab.—Persia. Cultivated in India.

Native synonyms:—

Aru, Rek, Chimnanu, Ghurghushtai, Mandala. *N.W.P.*

Peach tree gum was exhibited by Lieut. Hawkes at the Madras Exhibition of 1855.

Anacardium occidentale. *L.*

W. & Arn. Prod. 168.

Acajuba occidentalis, *Gaertn. Fruct.* t. 40 f. 2.

Cassuvium pomiferum, *Lam.*

Tree: leaves oval, rounded or slightly emarginate at the apex, narrower but obtuse at the base, bracteoles broadly ovate, acuminated; one stamen longer than the others, fruit sessile on the apex of the torus, reniform. (*W. & A.*)

Hab.—

Native synonyms:—

Kajee, Kaju, Hijli-badam, Cajoo. *H.*
Jidi-mamidi, Munta-mamidi. *Tel.*
Kola-mavah, Mundiri. *Tam.*
Kaju, Hijli-badam. *B.*
Jambu-monat. *My.*
Parunkimavah, Kajom-mara. *Mal.*
The-ho-thayet, Thubbamboo. *Burm.*

Cashew gum is usually in elongated stalactitic pieces or in cylindrical tears, varying in colour from a pale ochraceous white, almost colourless, to a dirty brown, rather dull and opaque, and not wholly soluble in water.

The Madras Jury Reports state that the trunk and branches of this tree yield, on being wounded during the ascent of the sap, a transparent gum similar in appearance to gum Arabic, for which it is a good substitute. This gum is sub-astringent, and being unpalatable to insects is particularly adapted for use where their depredations require to be guarded against.

The samples in the museum are from Shemooga and Travancore. "Ody" gum from the last-named locality appears to be the same thing. Whether the *Juay-yan* tree gum from Burmah is derived from the same source there is no evidence to show. It is apparently the same gum.

Cashew gum is only partially soluble in cold water, and partly swelling as in cherry-tree gum.

Armeniaca vulgaris. *P.*

DCand. Prod. ii. 532.

Prunus Armeniaca. *Roxb. Fl. Ind.* ii. 501.

Low tree of rather crooked growth; leaves broad, roundish, cordate or ovate, glandularly serrated; flowers white, tinged with dusky red. Fruit round, yellow within and without. Seed smooth, compressed.

Hab.—Armenia. Cultivated in India.

Native synonyms :—

Jaldaru, Chui, Barzha, Hari, Chir, Sari. *N.W.P.*

Dr. Stewart observes that a gum similar to gum Arabic exudes from wounds in this tree; and Mr. Baden Powell says that it is "an amber-coloured soluble gum."

Artocarpus incisa. *L.*

Desc. Fl. 8 t. 539. *Roxb. Fl. Ind.* iii. 527.

High tree; leaves roundish, pinnatifid, lobes oblong, pointed, male calyx bifid, style 2–3 partite.

Hab.—Pacific and East Indian Islands.

The native names of the bread fruit tree are—

Kebalsu. *Can.*
Nangka. *My.*

In Ceylon "Ratadel" is said to be the name given to the gum or resin from this tree.

Artocarpus integrifolia, *L.*

Tuss. Fl. 2 t. 4. *Roxb. Cor. Pl.* t. 250.

Tree; leaves elliptical, pointed, entire (rarely lobed), male calyx bipartite, style simple.

Hab.—Pacific and East Indian Islands.

Native synonyms—

Pila, Viram-pila, Peelah. *Tam.*
Panasa, Veru-panasa. *Tel.*
Kantal. *B.*
Phunus. *Dec.*
Pannas. *H.*
Fannas. *Mahr.*
Peing-nai, Pani-nai. *Burm.*
Halasoo, Alase-gana-mara. *Can.*
Sukun, Kluwi, Tambul, Nunka, Champada. *My.*
Kos, Heralee, Gediya, Wakara, Waela. *Cing.*
Tiun, Dahen. *N.W.P.*

The gum of the jack-tree was sent by Dr. Waring to the Exhibition at Madras in 1855 from Travancore. This gum is also collected in Ceylon under the name of "Kos." The Rev. J. Long says that the yellow viscid milky juice furnishes bird lime, leading to the inference that the so-called gum is a kind of gutta. A sample in the Museum Collection from Travancore is a black pitchy looking gum with a shining resinous fracture, resembling somewhat a very dark sample of Babool gum. It is a soluble gum.

Azadirachta Indica. *Juss.*

W. & Arn. Prod. i. 118; *Roxb. Fl. Ind.* ii. 394; *Wi. Ic.* t. 17.

Melia azadirachta, *Linn. DCand. Prod.* i. 622.

Tree: leaves simply pinnated, leaflets unequalsided, glabrous, serrated; panicles axillary, flowers small, white, fragrant.

Hab.—Peninsula. Hindostan. Assam.

The native synonyms are :—

Azad-darakhtehindi, Nib. *P.*
Ninb, Nimb. *H.*
Nim. *Duk.*
Bevina mara. *Can.*
Vembu, Veppam, Veppa marum. *Tam.*
Vepa chettu, Nimbamu. *Tel.*
Veppa, Ariya-veppa. *Mal.*
Nim, Nimgachh. *B.*
Nimba vrikshaha. *S.*
Limbacha-jhada. *Mah.*
Limbdanu jhada. *Guz.*
Kohumba, Nimba-gaha. *Guz.*
Tamabin, Thamakha, Thamaga, Kamakha, Khamakha. *Bur.*

Dr. Waring in his detailed account of the economic products of this tree makes no mention of its yielding a gum.

Dr. Birdwood says that it forms a portion of gum gattie and East India gum. Whilst O'Shaughnessy enumerates it as one of the gum producing trees. Samples were sent by Lieut. Hawkes and Dr. Cleghorn to the Madras Exhibition of 1855.

The museum sample from Bangalore is a clean bright amber coloured specimen in small tears and fragments of tears; that from Calcutta is in larger pieces and darker in colour, mixed with fragments of the well-known fibrous bark.

Barleria prionitis, *L.*

DCand. Prod. xi. 237; *Roxb. Fl. Ind.* iii. 36; *Wi. Ic.* t. 452.

Shrubby. The sterile spinous bracteas and bracteoles in four cleft fascicles, the fertile bracts subulate, spinous; leaves elliptic-oblong, attenuated at both ends, glabrous beneath, on the lines and margins slightly hairy; flowers sessile, axillary, verticelled, the terminal ones spicate; larger segments of the calyx ovate, spinously cuspidate, quite entire, glabrous; flowers bright orange. (*Drury*).

Hab.—Peninsula. Bengal. Sylhet. Assam.

Native synonyms of the tree—

Shem-moolie, Coletta-veetla. *Mal.*
Cantha-jathi. *B.*
Shem-moolie. *Tam.*
Mooloo-gorinta. *Tel.*

A sample of gum said to be derived from this source was sent from Madras to the Punjab Exhibition of 1864.

Bassia longifolia, *L.*

DCand. Prod. viii. 197; *Roxb. Fl. Ind.* ii. 523; *Wi. Ill.* t. 147.

Tree: leaves lanceolate, acuminate at both ends, petioles slightly villous; pedicels a half shorter than the leaf, sub-erect; corolla, 8–9 cleft; stamens, 16–20 in two rows; calyx rufous, pubescent; berry oblong, villous, yellow, size of a large plum, 3–1 seeded; flowers whitish. (*Drury.*)

Hab.—Southern parts of Coromandel Coast. Malabar.

Native synonyms—

Ippa, Pinna, Ippee. *Tel.*
Kat-illupi, Elupa or Elloopi. *Tam.*
Mohwa, Mohe. *H.*
Mahwa. *Guz.*
Ellupi. *Mal.*
Mee, Telmee. *Cyn.*
Madooka. *Sans.*
Moca. *B.*

The gum of Bassia longifolia, Eloopa gum, was contributed by Dr. Cleghorn to the Madras Exhibition of 1855. It is styled an "inferior" gum. By making deep incisions around the tree (and *B. latifolia*) a kind of white milky gum comes out, but of no use. (*Journ. Agri. Hort.* (1871) p. 231.)

Bauhinia parviflora. *Ham.*

Bauhinia emarginata. *Roxb.*

Bauhinia Vahlii. *W. and A.*

Bauhinia retusa. *Wall.*

Under the name of Sem-ke-gond is included the gum of more than one species of *Bauhinia*. It is said to be a brownish mild gum like that of the cherry tree. A sample very much of this character in the collection is referred to *Bauhinia emarginata*, it swells in water and behaves as a bassorine or cherry tree gum; a very small portion is soluble.

Borassus flabelliformis. *Linn.*

Linn. Fl. Zey. 395; *Roxb. Fl. Ind.* iii. 790; *Roxb. Cor.* t. 71, 72; *Martius Palm,* t. 108, 121, 162.

Stem, 30–100 feet; fronds terminal; leaves very large, fan-shaped, segments 2 cleft at the apex; petioles serrated and spinous on the margins; fruit large, spherical, obscurely brown when ripe. (*Drury.*)

Hab.—Malabar and Travancore, in sandy places towards the south. Coromandel. Bengal.

Native synonyms are—

Tal, Tari. *H.*
Tal, Tal-gach'h. *B.*
Panna-marum, Panang-kulloo. *Tam.*
Tatti-chettu, Karata-lamu, Potu-tadi Penty. *Tel.*
Tala. *Sans.*
Lontar. *My.*
Pana, Ampana, Carimpana. *Mal.*
Tal, Tal-gaha. *Cyng.*
Dom Tal. *Ar.*

Palmyra gum is described by Baden Powell as black, with a black and shining fracture. It was exhibited from Masulipatam at Madras in 1855.

Buchanania latifolia. *Roxb.*

Roxb. Fl. Ind. ii. 385; *W. & Arn. Prod.* i. 169.

Tree; leaves broadly oval or obovate, obtuse; branches of the panicles hirsute, with the flowers crowded; flowers small, greenish white. (*Drury.*)

Hab.—Mountainous parts of the coast. Belgaum. Salsette.

Native synonyms of the tree—

Chironji, Charvoli, Dhan. *H.*
Chinna-moral, Chara-chettu, Jaru-mamidi, Sara-puppoo, Chara-pappoo. *Tel.*
Katma, Kaat-mango, Moraeda. *Tam.*
Pia-sal. *Guz.*
Char. *Mahr.*
Chara. *Sans.*
Lumbo. *Burm.*

The Chironji gum is mentioned as one of the gums sent from Madras to the Punjab Exhibition of 1864. A sample was sent from Haidarabad for valuation, and the officiating conservator of forests reported that a good specimen of the tree yields about 5 lbs. of gum; it is not much exported, and has consequently no fixed market value. The reports here were "Inferior gum Arabic, part insoluble, and very foul, value "probably 25*s.* per cwt.;" another, "An inferior gum of Acacia type, fairly soluble, "and worth 15*s.* to 20*s.* per cwt."

This sample, above alluded to, was in irregular broken fragments, brittle, pale horn colour tinged with brown, tasteless, soluble in water, forming a colourless mucilage, but containing a portion of insoluble bassorin. The sample much mixed with impurities rendering the filtration of the mucilage necessary, otherwise the mucilage not inferior in strength to ordinary commercial Arabic gum.

In the proceedings Agri. Hort. Soc. of India (1840), April, p. 6, it is stated that the Peear yields a very hard adhesive gum of a clear white colour.

A report by Messrs. R. Scott Thomson & Co., on a sample of this gum is to the following effect:—"We have carefully examined the sample of the gum of "Buchanania latifolia, and find by chemical analysis it resembles the Bussorah gum, "but has superior adhesive properties similar to the gum Acacia, which gives it a "higher commercial value, as it can be used in the place of Acacia gum for dressing "cloth." (*Journ. Agri. Hort. Soc. Proc.* (1869) lxxxi.)

The Museum samples are the one alluded to above from Central India (4315), which is a picked sample, and one not much inferior not localised. Two samples from Nagode, one called "Goorga" gum, are dissimilar in many respects and somewhat doubtful.

Calyptranthes caryophyllœfolia, *Willd.*

Syzygium Jambolanum, *W. & A. Prod.* 1. 329; *DCand. Prod.* iii. 259.
Syzygium caryophyllifolium, *DCand. Prod.* iii. 270.
Eugenia Jambolana, *Roxb. Fl. Ind.* ii. 484; *Wi. Ill.* ii. 16.
Eugenia obtusifolia, *Roxb. Fl. Ind.* ii. 485.
Calyptranthes Jambolana, *Willd.*
Myrtus cumini, *Linn. Burm.* p. 115.

Arborescent; leaves oval or oblong, more or less acuminated or obtuse, feather-nerved, coriaceous; cymes panicled, lax, usually lateral on the former years' branches, occasionally axillary or terminal; calyx shortly turbinate, truncated; berry olive-shaped, often oblique; flowers small, white. (*Drury.*)

Hab.—Peninsula, Northern India.

Native synonyms—

Nawal, Neredi-chettu. *Tel.*
Navel, Naurei, Naridon kurra. *Tam.*
Madan. *Cyn.*
Jamoon. *H.*
Choto-jam. *B.*
Thabyehgah. *Burm.*
Naradidi, Vriksha. *Can.*

The Navel tree gum was sent from Masulipatam to the Madras Exhibition of 1855. Is it a kind of kino, as hinted by Dr. Birdwood, or is it a soluble gum? (vide ante).

Careya arborea, *Roxb.*

Roxb. Fl. Ind. ii. 638; *W. & Arn. Prod.* i. 334; *Roxb. Cor. Pl.* iii. 218; *Wight Illus.* t. 99, 100.

Tree; leaves obovate or oblong, slightly serrulate-toothed; flowers sessile, in short terminal few flowered spikes; fruit, broadly ovate; flowers large, greenish-white. (*Drury.*)

Hab.—Peninsula. Concans. Travancore. Foot of the Himalayas.

The native names of the trees are—

Buda-darmee, Kamba-karra, Dudippa, Kumbhi, Aray-chettu (?), Budada-nedi (?) *Tel.*
Kumbi, Cumbia. *Can.*
Bambonay, Bhaubhway, Baubwai. *Burm.*
Ban-bambhooai. *B.*
Kamba. *H.*
Koombha, Wae-koombha, Kombo. *Mahr.*
Peloa. *Mal.*
Kahatta, Kahata-gaha. *Cyn.*
Ave-mavo, Puta-tanni-maram, Pailœ-marum. *Tam.*
Kumbia, Luta-amaree. *Assamese.*

The sample exhibited at the Punjab Exhibition is described as a greenish gum.

Cathartocarpus fistula, *Pers.*

Cassia fistula. *W. & Ar. Prod.* i. 285.

Small tree; leaflets about 6 pair, broadly ovate, usually obtuse or retuse, sometimes slightly acuminated, glabrous, petioles without glands; racemes terminal, long, lax drooping; legumes cylindric, pendulus, with transverse partitions; cells 1 seeded, filled with pulp. Flowers bright yellow, fragrant. (*Drury.*)

Hab.—Peninsula. Bengal. Assam. Oude. Foot of the Himalayas.

Native synonyms—

Amaltas, Gurmilla. *H.*
Amultas. *Dec.*
Reyla, Suvarnam, Rella. *Tel.*
Konne, Koanay, Sarakonne. *Tam.*
Sondali, Sonalu, Banur-lati-gach'h. *B.*
Khyar-i-chembir. *P.*
Gurmalla. *Guz.*
Suvarnam, Suvarnuka. *Sans.*
Kakae. *Can.*
Chune, Mentus, Kakay. *Mal.*

Cassia fistula gum was contributed from Travancore to the Paris Exhibition of 1865. The only sample in the Museum is a dark coloured soluble gum from the same locality.

Cedrela toona, *Roxb.*

Roxb. Fl. Ind. i. 635; *Roxb. Cor. Pl.* 238; *W. & Arn. Prod.* i. 124.

Large tree, leaves abruptly pinnate, leaflets 6–12 pair, ovate-lanceolate, acuminated, entire or slightly toothed; panicles drooping; petals ciliated; sterile filaments none; ovules eight in each cell; capsule oblong; flowers small, whitish, fragrant. (*Drury.*)

Hab.—Peninsula. Foot of the Himalayas. Ravines at Khandalla.

Native synonyms—

Toona, Toon, Lood. *H.*
Tunna, Toon, Lood. *B.*
Malee, Wunjooli, Thoona-marum, *Tam.*
Thit-kadoe. *Burm.*
Tuna Cuveraca. *Sans.*
Suola-mara, Tunda. *Can.*
Toon, Kooruk, Thoon. *Mahr.*

Toon-ke-gond is enumerated by Dr. Wight as one of the gums of Coimbatore. It was included amongst the specimens exhibited by the late Dr. Royle in 1851. From the character of the timber one might suppose it rather a resin than a gum.

Celtis orientalis, *L.*

Wi. Icon. t. 602. *Roxb. Fl. Ind.* ii. 65.

Small tree, 15 feet, leaves alternate, bifarious, short-petioled, ovate-cordate, acuminated, minutely serrated, scabrous above, villous beneath; flowers axillary, aggregated on short two-cleft diverging peduncles; calyx five-parted; male and female flowers generally on a separate tree; drupe small, succulent, black when ripe, nut wrinkled, one-celled, one-seeded; flowers very small, green. (*Drury.*)

Hab.—Coromandel. Bengal. Travancore.

The native names given for the tree are—

Chakan-tubunna. *B.*
Mallam-toddali. *Mal.*

Drury says that this tree yields a gum resembling that of the cherry tree. (*Useful Plants.*)

Chickrassia tabularis, *Juss.*

Wi. Ill. t. 56.

Swietenia chickrassa. *Roxb. Fl. Ind.* ii. 399.

Leaves abruptly pinnated; leaflets 5–8 pair, obliquely ovate-oblong, unequal sided quite entire, hairy on the axils of the nerves beneath; panicles terminal, erect, flowers middle-sized, greenish-white. (*Drury.*)

Hab.—Cunnawady Hills near Dindigul.

Native synonyms—

Agley, Agle-marum. *Tam.*
Chickrassi. *B.*
Yimma, Nga-tai. *Burm.*
Pabba, Pubba. *Mahr.*
Dal-mara. *Can.*
Hulanhick-gala. *Cey.*
Dovedah. *Mal.*

A gum described as the produce of this tree was sent over in 1873 (No. 2055) from Madura. It consists of irregular tears, amber coloured or light brown, somewhat transparent but not brittle.

Citrus aurantium, *L.*

DCand. Prod. i. 539. *W. & A. Prod.* 97.

Citrus nobilis. *Lour.*

Leaves oval, elongated, acute, sometimes slightly toothed, petiole more or less dilated and winged; flowers white, large; fruit orange coloured, roundish or ovoid, usually depressed, rarely terminated by a small knob, rind with convex vesicles of oil; pulp sweet.

Hab.—Cultivated in India.

Native synonyms—

Narang, Naranj. *Ar.*
Kumla-nebu. *B.*
Orangen. *Dec.*
Naringi. *H.*
Naranj. *P.*
Naranga. *Sans.*
Kichidi. *Tel.*
Lieng-mau. *Burm.*
Jerooc, Simao-manis. *My.*
Kitchili, Kolinji-marum. *Tam.*

Orange tree gum is amongst the specimens mentioned in the catalogue of the Madras Exhibition of 1855 as sent from Masulipatam. It is said to be collected in the West Indies, and is included in the list of Jamaica gums.

Citrus decumana, *L.*

DCand. Prod. i. 539. *W. & A. Prod.* 97.

Young shoots pubescent; leaves oval-oblong, acute or obtuse, coriaceous; petioles long and much winged, flowers large, white, fruit pale yellow, rind with flat or convex vesicles of oil, pulpy vesicles separating from each other.

Hab.—Cultivated.

Chakotra, Batavineboo. *H.*
Bator-neboo, Batavineboo, *B.*
Bambalmas. *Tam.*
Edapundoo. *Tel.*
Bambelee-marunga. *Mal.*
Poomplemoos. *My.*
Maha-maram-jamboola, *Cyn.*

In 1855 Lieut. Hawkes sent to the Madras Exhibition a sample of Shaddock gum.

Citrus limetta, *Risso.*

W. & Arn. Prod. 98. *DCand. Prod.* i. 539.

Leaves oval or oblong, often toothed; petiole more or less winged or margined; flowers small, white; fruit pale yellow, ovoid or roundish, terminated by a knob; rind with concave vesicles of oil, pulp watery, sweetish, or insipid, or slightly bitter.

Hab.—Cultivated.

Nemboo, Meeta-neemboo. *H.*
Heboo, Neboo. *B.*
Elemitchum. *Tam.*
Nemma-pandoo, Gajanimma. *Tel.*
Eroomitchee narracum. *Mal.*
Dehi. *Cyn.*

Lime tree gum, sent from Masulipatam to the Madras Exhibition of 1855, is almost the only record of this as an Indian gum tree.

Citrus medica, *L.*

Wi. & Arn. Prod. 98. *DCand. Prod.* i. 539.

Young branches rigid; leaves oblong, pointed; petioles simple; flowers white, tinged with red; fruit obovoid, deeply furrowed and wrinkled; terminated by a knob; pulp very slightly acid.

Hab.—Cultivated.

Native synonyms—

Toorunj. *P.*	Lungamoo, Bijapuramu, Madi-phulla-chettu, Dubba-chettu. *Tel.*
Leemoo. *H.*	Jarok. *My.*
Beg-poora. *B.*	Ooturinj, Utruj. *Ar.*
Sedaran. *Cyng.*	

Citron tree gum is named as one of the gums sent from Masulipatam to the Madras Exhibition of 1855, but no remark is made as to its character or quality, and it is presumed that it had interest rather as a curiosity than as being a staple product.

Cocos nucifera, *L.*

Linn. Fl. Zey. Roxb. Fl. Ind. iii. 614. *Roxb. Cor.* i. 73. *Mart. Palm.* 62, 65, 88.

Stem tall, flexuose, thickened at the base; fronds spreading; pinnæ linear-lanceolate, acuminate; female flowers sub-globose; drupes very large, ovate, three cornered. (*Drury.*)

Hab.—Common on all shores of the Peninsula.

Native synonyms are—

Naril, Narel. *H.*	Pol-nawasi, Tæmbili. *Cyn.*
Narikel. *B.*	Nur. *My.*
Taynga. *Tam.*	Jowzhindee, Nardjil. *Ar.*
Tenga. *Mal.*	Tenkaia, Narikadam, Kobbari. *Tel.*

Coco-palm gum was sent to the Madras Exhibition of 1855 from Travancore.

Conocarpus latifolius, *Roxb.*

Roxb. Fl. Ind. ii. 442. *W. & A. Prod.* i. 316. *Wi. Icon.* t. 994.

Anogeissus latifolius, *Wall.*

Large tree, with white bark, leaves eliptical, or obovate, obtuse or emarginate, smooth; peduncles branched, bearing several heads of flowers, fruit imbricated upwards, produced at the apex into a long beak; flowers small, pale, greenish. (*Drury.*)

Hab.—Islamabad. Kennery jungles. Valleys of Concan rivers. Inland Deccan hills.

The native synonyms are—

Tella-neredu, Sheri-manu, Veckale, Duca. *Tel.*	Dinduga. *Can.*
Vellay-naga, Veckalie, Tella-narudoo. *Tam.*	Dhac, Chal, Kuladhan. *N. W. P.*
Thoura, Dhowra. *Mahr. H.*	

"The gum, which exudes from incisions made in the bark," writes Stewart, "is used "in cloth printing." The officiating Conservator of Forests, in sending over a sample from Haidarabad for valuation, remarks that a good specimen of the tree yields about four pounds of gum, and that it sells at the rate of two annas a seer, or about 140 rupees per ton. The trees are plentiful in the Melghât.

The Assistant Secretary to the Chief Commissioner of the Central Provinces, reports when sending over sample of this gum that some 4,500 tons of the better kinds of gums could readily be supplied on the Great Indian Peninsula Railway at from 100 to 200 rupees a ton. The "dhoura" and "badul" are said to yield the best gum. Also good gum is procured from the Char, Ber, Palas, Saj, Ain, Tendu, Khair, &c. There are numberless other gum bearing trees, the gum of most of which is not of any market value. All these gums are readily procurable in all districts in large quantities. But the different kinds are brought to market so mixed up together that it is impossible to distinguish between them or make a guess at the amounts of the different sorts that could be supplied. Gums are generally collected merely for local consumption. Small quantities are, however, exported into the Berars; from Sagar to the north, and from the Narbada division towards Bombay. During the last year or so two Bombay European firms have sent agents into the interior of the Nimar district to make arrangements for a permanent supply. They have not as yet been very successful. In some parts the Gonds gain a livelihood by gathering gum, but the price is so low that in most districts men find the cutting of grass and sticks much more profitable. As a rule women and children are employed in the work. In Nimar their wages are the weight of the gum collected in wheat or double its weight in jowari. A large demand at higher prices than those now offered would not only greatly increase the out-turn, but bring into the market a far superior article. At present, as has been stated, all kinds are gathered promiscuously and offered for sale as gum, not as the product of any particular tree, except, indeed, occasionally when the gatherers are under contract to supply the Doura or some other gum for medicine. But these pure gums are procurable in very small quantities. Not only this but the gum offered for sale always contains a quantity of sand, leaves, and bark. No attempt is made to keep it clean. It is gathered carelessly before it is sufficiently dry to come away by itself, and in consequence pieces of bark are torn off with it. But it may well be questioned whether this industry would ever become a remunerative one in these provinces so far removed from the sea. Gum bearing trees grow as extensively on the seaboard, where facilities for export are much greater.

Of the samples of this gum in the collection, that from Salem (1873) is in more or less irregular tears, adhering to fragments of bark, clear, transparent, pale honey coloured, more or less brownish, tough and not friable, tasteless, for the great part soluble in cold water.

A picked sample (No. 5) from the Central Provinces is in elongated cylindrical or vermiform tears, half an inch in diameter, sometimes three times as long, and curved, clear, transparent, pale horn colour, or almost colourless, forming a clear colourless mucilage with cold water, but with a small proportion of bassorin swelling and not dissolving. The mucilage was quite colourless, as tenacious as ordinary gum arabic, but must be poured off from the less soluble portion, which sinks to the bottom. The four reports on an unpicked, ordinary sample of this gum were: (1) inferior gum arabic, much too drossy, very weak gum. It should be picked clean. Sample of little value, probably 15*s.* per cwt. (2) Arabic ghatty gum, 22*s.* 6*d.* to 25*s.* per cwt. (3) Rather low gum of Arabic or Acacia kind, and partial solubility, value about 10*s.* per cwt. (4) Mucilaginous gum, very weak, and contains, say, 50 per cent. insoluble matter. The sample contains a large quantity of bark mixed with the gum. A clean sample might obtain, say, 15*s.* per cwt., but very unsaleable.

The valuations in this case varying from 10*s.* to 25*s.* per cwt. might at first sight seem extraordinary and inexplicable. It must, however, be borne in mind that the article submitted was entirely new to the market, and had no real commercial value. Brokers can calculate the value of recognised produce in the market at any time by comparison of qualities, and the relations between supply and demand, but when a new product is submitted to them for valuation, all determination on fixed bases vanish, and their estimate can only be based on individual opinions as to what might be the probable status it would occupy in the market. Therefore it is that all valuations of unknown and untried substances must be unsatisfactory, and will in most cases fall short of the real value obtained when the product is offered in bulk. In the present instance the gum could not have been rated as a gum arabic, except as a very low quality, because it had not been tested, and had not acquired a reputation or a *locus standi.* Valuations of this kind should be accepted, therefore, at what they are worth, and not invested with a fictitious importance. Neither can they at the same time be open to the charge of inconsistency, when, as in this instance, opinions differ so widely, inasmuch as they are opinions only, as to the probable risk which purchasers might be induced to take, or the probable effect of supply in originating a demand.

The specimens in the Museum Collection are, the above picked white gum from Central Provinces, a darker and more mixed sample from Bangalore, and another from Shemooga. Two specimens of gums sent from Baroda to Vienna in 1873, and called respectively White Sadada and Black Sadada are mixed, containing evidently a portion of this gum. The same observation applies also to a Dhawa gum from Nagode, and one under the same name from Chutterpore.

Cordia Rothii, *Rœm.*

Rœm. and Schul. Sys. iv. 798. *DCand. Prod.* ix. 480.
Cordia reticulata, *Roth.*
Cordia angustifolia, *Roxb. Fl. Ind.* i. 595. *Wi. Ic.* 1379.

Small tree; leaves sub-opposite, lanceolate, obtuse, mucronate, entire, rough; panicles terminal and lateral, rather shorter than the leaf, many flowered; flowers small, white; fruit size of a pea, orange-yellow. (*Drury.*)

Hab.—Mysore. Western Coasts.

Gund, Gundni. *H.*	Narvallie-marum. *Tam.*
Nakkeru, Chinna-botku. *Tel.*	Bokur. *Mahr.*

The gum of this tree is enumerated by Dr. Wight amongst the produce of Coimbatore.

Elate sylvestris, *Linn.*

Phœnix sylvestris. *Roxb. Fl. Ind.* iii. 787.

Stem long, thickish; pinnæ arranged in sub-opposite fascicles, ensiform, terminated by a spine, tetrastichous.

Hab.—Common throughout the country.

Seyndie. *H.*	Eetchumpannay. *Tam.*
Khajoor. *B.*	Khurjjooree. *Sans.*
Eeta. *Tel.*	Date palm. *Eng.*

Baden Powell describes this as a glossy dark brown gum, soft enough to be flexible. It was sent from Masulipatam to the Madras Exhibition of 1855.

Elœodendron paniculatum, *W. & A.*

Wi. & Arn. Prod. 157.

Leaves elliptical, green, bluntly serrated, on long petioles; panicles dichotomous, without a flower in the fork, about as long as the leaves, ultimate pedicels ternate. (*W. & A.*)

Hab.—

Native names—

Jumrasi. *H.*

A sample of *Jumrasi* gum was sent over for valuation from Central Provinces in 1873, which was referred by Capt. Doveton, Conservator of Forests, to this source. Whilst doubting the accuracy of this determination, it must be conceded that the gum is worthy of attention. It is in roundish tears of variable size, the majority not exceeding half an inch in diameter, externally finely rugose, and minutely cracked, with a shining fracture, rather brittle, some tears almost colourless, others dark reddish brown, with intermediate shades of amber and brown. Tasteless. Soluble in water forming a tenacious mucilage of a sherry colour.

Emblica officinalis, *Gærtn.*

Phyllanthus emblica, *Linn. sp.* 1393; *Roxb. Fl. Ind.* iii. 671.

Phyllanthus glomeratus, *Wall.*

Tree, leaves densely distichous, narrow linear-elliptic, obtuse at both ends, or a little acute at the apex, recurved at the margin, glabrous; petioles very short; flowers very numerous fascicled, males pedicelled, females sessile; glands of the male flowers very often none; urceolus hypogynous, cylindric, with many small divisions at the apex; anthers shortly apiculated; ovary ovoid; styles large, compressed above, twice dichotomously divided, papillose, connate at the base into a column nearly equalling the ovary; fruit globose, large, sub-baccate, fleshy, slowly opening as a capsule.

Hab.—Common in most parts of India.

The native synonyms are—

Usri, Usirika-manu, Amla-kamu, Oosarika, Oosarica, Usrika, Usirika, Assereki. *Tel.*	Kadondong, Malaca. *My.*
	Nelli. *Mal.*
Nellekai, Nellikai, Nelli-maram, Topa-nelli, Nelly-marum, Nelle-pale, Nelu. *Tam.*	Amusada-nelli, Nelli-gass. *Cyn.*
	Amluj. *A.*
Amla, Amlika, Amliki, Anola, Aonla, Aruli, Aungra, Aoula. *H.*	Aonla, Aonli. *Dec.*
	Aonli. *Mahr.*
Anola, Aonla. *B.*	Ohalu, Gondhona, Olla-awla. *Uriya.*
Amalaca, Amlaki, Umriti, Amlaka. *Sans.*	Mala-ka, Yee-pye. *Burm.*
Nelli-mara, Nilika mara, Toopoonellee, Nelly-mara. *Can.*	Amlokhi. *Ass.*
	Nellee. *Can.*

This gum was sent from Masulipatam to the Madras Exhibition of 1855. The sample referred to this source in the collection is rather brittle, in dark stalactitic pieces adhering to bark sent from Salem to Vienna.

Cicca disticha, L., is also credited with producing gum, but whether it does so or has been confounded with this tree is open to inquiry. A sample under the name of *Cicca* is in the museum collection from Shemooga.

Erythrina Indica, *W. & A.*

Wi. & Arn. Prod. i. 260. *DCand. Prod.* ii. 412. *Roxb. Fl. Ind.* iii. 249. *Wi. Icon.* 51.

Tree, armed with black prickles; stipules falcate; petioles and leaves unarmed; terminal leaflet broadly cordate; calyx spathaceous, contracted and 5-toothed at the apex; vexillum three times shorter than the calyx, many times longer than the wings; keel about the length of the wings, with its petals distinct; stamens monadelphous, with the sheath entire at the base, thence diadelphous with the tube split; racemes terminal, horizontal; legume 6–8 seeded; flowers large, dark purple scarlet. (*Drury.*)

Hab.—Coromandel. Concans. Assam.

Madagu, Mahameda, Badapuchettu, Badida-chettu? Badi-dapu-chettu, Barijamu, Barja-pu-chettu, Barjupu-chettu, Barcha-napa, Badida-chettoo, Badidil-chettu, Badida-karra, Bandita-chettu. *Tel.*
Kaliana-murukai, Murukka-maram, Muluku-murukka, Moor-koo-marum, Moorookoo-ma-rum, Muruka-marum, Moochoo-maram, Mu-ruku. *Tam.*
Palita-mandar, Palta-mandar. *B.*
Moduga-vriksha, Pangara. *Can.*
Furrud, Pangra, Panjirah, Palta-mandar, Paun-jerah. *H.*
Pangara, Panjirah. *Mahr.*
Erabadoo-gass, Erabadu-gaha. *Cyn.*
Chalo-dhona? Chaldua. *Uriya.*
Kashy. *Burm.*
Mundara. *Sans.*
Pangree. *Bombay.*
Mooloo-moorikah. *Mal.*

A small sample of one ounce of this gum was sent over from Madras to the London Exhibition of 1862. It was a dark, pitchy brown, opaque substance of no promise, much more resembling, to the eye, fragments of Pegu cutch than gum.

Feronia elephantum, *W. & A.*

W. & Arn. Prod. i. 96. *Wight Icon.* t. 15. *Roxb. Fl. Ind.* ii. 411.

Large tree; leaves pinnated, leaflets 5–7, almost sessile, obovate, petioles slightly winged; racemes lax, axillary and terminal; style scarcely any, flowers small, pale pink, with crimson anthers, fruit about the size of an apple, with a hard greyish rind, seeds immersed in fleshy pulp.

Hab.—Coromandel. Western Coast. Guzerat.

The following native names are given in the supplement to the Pharmacopœia of India to the gum of this tree—

Samaghe-kabit. *Ar. P.*
Katbel-ki-gond. *H.*
Vilam-pishin. *Tam.*
Velaga-banka. *Tel.*
Vilam-pasha. *Mal.*
Byalada-gondu. *Can.*
Kothbel-gun, Katbel-gun. *B.*
Kapidtha-niryasam. *Sans.*
Kavita-gonda. *Mah.*
Kavit-gon. *Guz.*
Divul-melliyam. *Cing.*
Ti-si. *Burm.*

Ainslie says that the Wood Apple gum is used by the dyers and painters, particularly the miniature and chintz painters; it is also employed in making ink and certain varnishes, and by the bricklayers in preparing a fine kind of whitewash. He also says that there are several trees which afford gums which resemble the true gum Arabic, but perhaps the produce of none of them comes nearer to it than this, which is commonly used for medicinal purposes by the Vijtians in these provinces.

According to the Pharmacopœia of India, this gum occurs in the form of irregular, semi-transparent reddish brown tears. Treated with water it affords a brownish tasteless mucilage, not less adhesive than that of gum Arabic.

The Madras Jury Reports observe that Wood Apple gum is very abundant, and with Babool gum yields the well known East Indian gum Arabic. Mr. Rohde mentions that from its ready solubility without residue it gives the best mucilage for making black ink.

The specimens in the Museum Collection are, one scarcely distinguishable from Arabic gum from South Arcot, one not wholly soluble in large tears from Madras, a black and very doubtful specimen from Travancore, and a gum of good quality from Burmah, referred doubtfully to a species of Acacia, but which seems rather to belong to this tree.

Garuga pinnata, *Roxb.*

Roxb. Fl. Ind. ii. 400. *W. & A. Prod.* i. 175. *Wi. Ic.* t. 1594–5.

Large tree; leaves unequally pinnated, deciduous; leaflets nearly sessile, crenate-serrated; sepals erect; disc fleshy, 5-cleft; style about as long as the petals; drupe globose, fleshy, with from 1 to 5 one-seeded nuts: flowers panicled, yellowish-white, often covered with a mealy kind of substance. (*Drury.*)

Hab.—The Ghauts. Coromandel. Assam. Northern India.

Garugu, Garuga-chettu. *Tel.*	Kuruk, Kharpat. *H.*
Karre-vembu, Carri-vembu-maram. *Tam.*	Koorak. *Mahr.*
Khyong-yook, Khyong-youk. *Burm.*	Mohi, Moea. *Uriya.*

Curvamboo gum, the produce of this tree, exhibited by Dr. Cleghorn at the Madras Exhibition of 1855, is described as an "inferior" gum.

Gossypium herbaceum, *L.*

DCand. Prod. i. 456. *Hook. Fl. Ind.* i. 346.

Annual or perennial, hairy or sub-glabrous, leaf-lobes broadly ovate acuminate; flowers yellow with a purple centre, rarely wholly yellow or white or purple, petals spreading, bracteoles not divided below the middle, sometimes entire or nearly so; cotton white or brownish, adherent to the seeds, overlying a grey or greenish down. (*Hooker.*)

Hab.—Cultivated in India.

Kapas, Rooi. *H.*	Capas. *My.*
Kapase, Tula. *B.*	Kootn, Beersoon. *A.*
Pungie, Paratie, Van-paratie. *Tam.*	Poombeh. *P.*
Puttie. *Tel.*	Cay-haung, Hoa-mien. *Chin.*
Kapas. *Dec.*	Kobung. *Mongolia.*
Kapu. *Cey.*	

Cotton tree gum was sent by Dr. Waring to the Madras Exhibition of 1855, from Travancore, probably only as a curiosity.

Grislea tomentosa, *Roxb.*

Roxb. Fl. Ind. ii. 233. *Bot. Mag.* t. 1906. *Roxb. Cor.* i. t. 31.

Lythrum fruticosum, *Linn.*

Shrub; branchlets pubescent; leaves opposite, quite entire, lanceolate, somewhat cordate at the base, sessile, hoary underneath; petals usually 6, stamens declinate, usually 12; capsule oblong; peduncles axillary, several flowered; flowers middle-sized, scarlet, with rose-coloured petals. (*Drury.*)

Hab.—Peninsula. Concans. West Ghauts. North India.

Names of the tree—

Dhauri, Dhub, Dhanga-phul, Dhaephool. *B.*
Sirinjie. *Tel.*
Dhaitee, Dhauree. *Mahr.*

The gum called dhaura or dhau-ka-gond is brought from Mewar and Harowtee, and is abundant, white in colour, like the katira and tragacanth gums, swells in water; in dyeing cloth it is applied to those parts that the dye is not wished to touch; it is eaten in "luddoo," one maund costs 10 rupees. (*Balfour.*)

The Museum samples do not all agree in character. One sample from Allahabad is a good strong gum in tears. A sample from Bengal is in smaller fragments. One from Ahmedabad appears to be mixed, and one from Indore is very much like the gum of *Conocarpus.* The "Dhokra" gum from Goonah is in rounded tears about the size of a filbert, and may belong to this species.

Gyrocarpus Jacquinii, *Roxb.*

Roxb. Cor. t. 1. *Roxb. Fl. Ind.* i. 445.

Gyrocarpus Asiaticus. *Willd.*

Large tree; leaves crowded about the extremities of the branchlets, broad, cordate, three-nerved, often slightly lobed, above smooth, below downy, with two pits on the upper side of the base; petioles downy, panicles terminal, divisions two-forked, hermaphrodite flowers solitary, sessile in the division of the panicle; calyx five-sepalled, segments unequal, interior pairs large, wedge-shaped, three toothed, expanding into two long membranaceous wings, flowers small, yellow; capsule globular, wrinkled, one celled, one valved, size of a cherry, ending in two long lanceolate membranaceous wings.

Hab.—Coromandel.

Zaitun. *H.*
Tanuku, Kumar-pulki, Punki, Poonkee. *Tel.*

The Poongallee gum sent to the Punjab in 1864 from Madras is referred doubtfully to this source. The gum is described as black and like dried tar.

Heritiera littoralis, *Dryand.*

DCand. Prod. 1, 484. *Roxb. Fl. Ind.* iii. 142.

Tree; leaves oval, rounded at the base; panicles axillary, flowers small, red.

Hab.—Peninsula.

Doengngoen-kontol. *My.*
Ka-na-zoe, Kon-zo-za-loo, Kunna-zoo, Kannatso. *Burm.*

This gum or resin is mentioned as one of the products of Ceylon under the name of "Mendora." Whether it is procured in India or in Burmah, we have no evidence to show.

Jatropha curcas, *L.*

Roxb. Fl. Ind. iii. 686.

Small tree or shrub; leaves scattered, broad cordate, 5-angled, smooth, panicles terminal, or from the exterior axils, cymose, many flowered; male flowers at the extremities of the ramification on short articulated pedicels, the female ones in their divisions, with pedicels not articulated; calyx 5 leaved, corolla 5 petalled, campanulate, somewhat hairy, styles 3, short; flowers small, green; ovary oblong, smooth.

Hab.—Coromandel. Travancore.

Bag-bherindha. *H.*
Bagh-dharanda, Bag-bherenda. *B.*
Caat-amunak, Caar-noochie. *Tam.*
Nepalam, Adivicamida. *Tel.*
Erundi. *Dec.*
Caak-avenako. *Mal.*
Mara-narulle. *Can.*
Rataendaroo. *Cey.*

Gum from this plant occurs in the Madras Catalogue of 1855 as an exhibit from Masulipatam.

Melia Azedarach, *L.*

W. & Arn. Prod. i. 117. *Wight Icon.* t. 160.

Leaves deciduous; leaflets about five together, obliquely ovate-lanceolate, serrated, finely acuminated; petals nearly glabrous; peduncles axillary, panicled above, many flowered; flowers smallish, white, externally lilac at the top, fragrant.

Hab.—Northern India. Deccan. Concans.

Native synonyms for the tree are—

Bakayan, Maha-ninb. *H.*
Gouri-nim, Gouli-nim. *Dec.*
Malai-vembu, Malai-veppam. *Tam.*
Konda-vepa, Turaka-vepa. *Tel.*
Bottada-bevina. *Can.*
Maha-nim. *B.*
Parvata-nimba-vrikshaha. *S.*
Dongra-cha-limba-cha-jhada. *Mah.*
Maha-nimba. *Cey.*
Simbo-tama-bin, Simbo-thama-kha, Simbo-thamaga, Simbo-khamakha. *Bur.*

A specimen of gum in the collection, said to be derived from this tree, and received from Salem, is in irregular tears, rather adhesive and dull, with a shining fracture, amber coloured or brownish, rather friable, mixed with fragments of bark; tasteless; soluble in water. The gum of this tree was one of the contributions of Dr. Cleghorn to the Madras Exhibition of 1855.

The only museum specimen (No. 2888) was received from Salem, and is a brown soluble gum in large tears, not unlike in appearance to wood-apple gum.

Melia sempervirens, *L.*

Roxb. Fl. Ind. ii. 305. *W. & A. Prod.* 1. 117.

Leaves evergreen; leaflets ovate, sometimes cordate at the base, cut or serrated, with taper entire apices, glabrous; petals nearly glabrous.

Hab.—Cultivated.

Bukarjun, Bukayun. *H.*
Maha-nimba. *B.*
Ban, Bam ? *A.*
Bukayun, Bukain, Daracht-i-azad. *P.*
Bukaena ? *Parb.*
Bakai ? *New.*
Bakain, Drek. *Punj.*

It is named amongst the gums of Jamaica, and, though a West Indian plant, is found in India, where it seems to be little known as a gum producing plant. Baden Powell remarks that the gum looks like Mochras (*Punj. Prod.* p. 412), and, therefore, it is probably of little importance.

Michelia Champaca, *L.*

Roxb. Fl. Ind. ii. 656; *W. & A. Prod.* 6.

Tree; leaves lanceolate, acuminated, glabrous, flowers on short peduncles; spatha of one leaf, anthers obtuse.

Hab.—

Champakamu, Champeyamu, Gand'ha-p'hali, Hem-an-gamu, Chempaka-manu, Heman-push-pakamu, Kanchanamu, Sampenga-chettu, Kanchanapu-chettu ? *Tel.*
Champa, Champaka. *B.*
Sumpaghy, Uru-sampige. *Can.*
Chumpa. *Dec.*
Bongas-jampacca, Mangliet. *My.*
Champakam, Schampakam, Siapangam. *Mal.*
Champaca. *Sans.*
Konchona ? Champā. *Uriya.*
Changal, Paranyan. *Burm.*
Phul-sopa, Fool-sopa, Fool-chopa. *Ass.*

Champaca Pal occurs in the Madras Exhibition Catalogue of 1855, as exhibited by Lieut. Hawkes.

Mimusops Elengi, *Linn.*

DCand. Prod. viii. 202; *Roxb. Fl. Ind.* ii. 236; *Roxb. Cor.* t. 14; *Wi. Icon.* t. 1586.

Tree; leaves elliptic oblong, obtusely acuminate, smooth; fascicles of pedicels axillary, shorter than the petiole, 3–6 flowered; pedicels rusty pubescent, lobes of the calyx lanceolate acuminate, equal, four outer ones rusty-velvety outside, smooth within; four inner ones narrower, whitish-velvety outside, smooth within; flowers middle sized, white, fragrant. (*Drury.*)

Hab.—Circars. West Ghauts. Peninsula. Sylhet.

Magadam, Magludam-maram, Mayugadi-maram, Maghidam. *Tam.*
Pogada, Pogada-manu, Pogada-mannoo, Pagadupu. *Tel.*
Mugali, Mugali-mara, Minjulu ? Moogali-mara. *Can.*
Mulsari, Bakula, Mulsari-ka-jhar. *H.*
Kya-ya. *Burm.*
Bakula. *B.*
Bacul, Bholsari, Mulsari, Taindu. *Dec.*
Bakul. *Mahr.*
Elengi. *Mal.*
Bakula. *Sans.*
Moone-malgass. *Cyn.*
Bokul. *Ass.*

Dr. Birdwood observes that this tree yields the Pogada gum of the Madras Presidency.

Mimusops Kaki, *W.*

Roxb. Fl. Ind. ii. 238.

Tree; leaves alternate, obovate, very blunt; silvery or hoary beneath, crowded at the ends of the branches; flowers fascicled, hexandrous; fruit oval, drooping; flowers yellowish white, tinged with rose.

Hab.—Malabar.

Kirni, Khirni. *H.*
Manil-kara. *Mal.*
Kheerokolee. *Uriya.*

According to Dr. Birdwood this tree also affords a gum.

Morus Indica, *Willd.*

Roxb. Fl. Ind. iii. 596.

Diœcious, subarboreous. Leaves ovate, cordate, long, taper pointed, serrate, smooth, aments oval. Styles single, half two cleft.

Hab.—

The native synonyms are—

Toot. *B.*
Toola. *Sans.*
Tulklu, Tuti, Tut, Shah-tut. *N. W. P.*

Mulberry tree gum was amongst the samples exhibited by Lieut. Hawkes at the Madras Exhibition of 1855.

Nerium suaveolens.

Baden Powell in his Punjab products mentions a dull red gum, said to be derived from *Nerium suaveolens* sent from Madras to the Punjab Exhibition of 1864.

Odina Wodier, *W. & A.*

W. & Arn. Prod. i. 171. *Roxb. Fl. Ind.* ii. 293. *Wight Icon.* t. 60. *Royle Illus.* t. 31, f. 2.

Large tree, leaves alternate, unequally pinnated, leaflets 3–4 pair, almost sessile, oblong ovate, acuminated, quite entire, sepals rounded; stigmas simple; drupe very hard, 1-celled, 1-seeded; racemes terminal, fascicled; flowers small, greenish-yellow, purplish externally. (*Drury.*)

Hab.—Coromandel mountains. Travancore.

The native synonyms already ascertained are, for the gum—

Jhingun-ke-gond, Kenni-ke-gond, Jingna-gond, Kana-gond. *N. W. P.*
Ging. *Cyng.*

And for the tree—

Kiamil, Kambal, Batrin, Kemal, Dila. *N. W. P.*
Cushmulla, Jingam, Kashmala. *H.*
Jiwul, Jival. *B.*
Woodian, Odiya-maram. *Tam.*
Waddi-gampina, Oddi-manu. *Tel.*
Woodier, Udi-maram. *Mal.*

Dr. J. L. Stewart writes of this gum that "Large quantities exude spontaneously "from the bark of this tree (at Bijnour), and much is collected for export. It is used "by dyers, cloth printers, for making ink, and in medicine, and is practically "divided into two kinds. These are, 1st, the white, generally called *kanne*, of which "a man can gather one seer a day, to be sold to the dealers; and, 2nd, the black "*jingan ka gond*, which consists nominally of what has fallen to the ground, but is "almost invariably mixed and sold with that of Conocarpus. Of the black gum one "man can collect five seers in a day. Both are gathered about April." (*Agri. Hort. Journ.* xiii. 311.)

It has already been remarked that great confusion appears to exist as to the produce of *Odina wodier.* Dr. Royle and O'Shaughnessy, and probably also Dr. Ainslie, were of opinion that a resin was produced by *Odina*, whereas recent writers, as Dr. Stewart Baden Powell, &c., adhere to the opinion that it is a soluble gum. In all instances, except one, the gums sent over to European exhibitions from India and to the India Office for report have been soluble gums. It is true that Dr. Royle, in his "Illustrations," remarks that a gum resin exudes from the bark of Odina Wodier, which is so abundant as to form an article of commerce; it is called in Northern India, Jhingun-ke-gond, and Kenni-ke-gond. These native names are singularly confirmed by Dr. Stewart, as above cited, but whilst Royle states that it is a gum resin, Stewart evidently has a soluble gum in view in his remarks.

Baden Powell describes a specimen of this gum in cylinders, which had been dissolved and run into a bamboo, after the manner of gamboge, in order again to solidify, and in this artificial form it was exhibited at the Punjab Exhibition.

A sample of Mohin gum, supposed to have been derived from this source, was sent over here for valuation from Haidarabad. The officiating Conservator of Forests states that a good specimen of the tree yields five pounds of gum, that it is not much exported, and consequently has no fixed market value. The reports on this gum were (1), Gum arabic, similar character to chirongee, but more inferior, and too low for this market; (2), Arabic hoggy, mixed, 25*s.* to 30*s.* per cwt.; (3), similar to chirongee, but more lumpy, about 10*s.* to 15*s.* per cwt.; (4), Mucilaginous gum, containing a large quantity of viscid insoluble matter, thereby rendering it nearly useless, as it would be impossible to separate the soluble from the insoluble gum. If brought on the market it would most probably be bought and shipped abroad for the purposes of adulteration. It has really little or no intrinsic value.

Of the gums in the collection attributed to this source, is one which has been in store for very many years, and consists of small broken fragments of tears, seldom larger than a pea, mixed with dusty gum resembling coarse sand. It is of a very light colour, and entirely soluble, forming a strong almost colourless mucilage.

Another sample is in much larger tears, but similar in colour; it is not so soluble, but forms a good mucilage. It has the appearance of being mixed with some less soluble gum, probably that of *Conocarpus*. This is the sample alluded to from Haidarabad.

A specimen from Simli pass is similar to the first named, but darker. Whereas " Koonco-gond" from Calcutta is very dark and doubtful. A sample from Chingleput referred to this source is a resin.

Poinciana elata, *W. & A.*

W. & Arn. Prod. i. 282. *DC. Prod.* ii. 484. *Roxb. Fl. Ind.* ii. 355.

Cæsalpinia elata, *Sw.*

Tree unarmed, leaflets linear, obtuse; calyx more or less pubescent, particularly inside; sepals coriaceous, equal, lanceolate, acute; æstivatian valvular; petals fringed; ovary villous; flowers large, yellow. (*Drury.*)

Hab.—Coromandel. Malabar. Guzerat. Arcot.

Native synonyms—

Pade-narrayan. *Tam.*
Sooncaishla, Sunkaser. *Tel.*
Neerangi. *Can.*

A small sample of this gum was sent to the London Exhibition of 1862 by Dr. Shortt, of Chingleput.

Poinciana regia, *Bojer.*

Bot. Mag. t. 2884. *Voigt. Cat.* 245.

Hab.—Madagascar. Cultivated in India.

This is not a native tree of India, but Dr. Birdwood observes that it flourishes luxuriantly in Bombay, and yields gum abundantly.

Pongamia glabra, *Vent.*

W. & Arn. Prod. i. 262. *DCand. Prod.* ii. 416. *Wi. Ic.* t. 59.

Galedupa Indica, *Roxb. Fl. Ind.* iii. 239.

Dalbergia arborea. *Willd.*

Tree. Leaflets 2–3 pair, ovate or obovate, acuminated, glabrous; vexillum, with two callosities at the base of the limb and decurrent along the claw; racemes axillary, many flowered, about half the length of the leaves; pedicels in pairs; legume oblong, nearly sessile, thick, with a short recurved beak, tumid along both sutures; flowers middle-sized, bright purple.

Hab.—Coromandel. Concans. Deccan. Patna. Assam.

The synonyms of the tree are—

Kurung, Caranj, Karunje. *H.*
Kurunja. *B.*
Poongu-marum. *Tam.*
Kanoogoo, Kanoogamanoo. *Tel.*
Pongam. *Mal.*
Canaga. *Can.*
Mokul-karanda. *Cing.*

The specimen sent to the Punjab from Madras in 1864 is described as a thick black untransparent exudation.

Prosopis spicigera, *L.*

W. & Arn. Prod. i. 271; *Rox. Cor. Pl.* i. t. 63.

Adenanthera aculeata. *Roxb.*

Somewhat arboreous, armed with scattered prickles or sometimes unarmed; leaves rarely simply pinnated, usually bipinnated with 1–2 pair of pinnæ, and a gland between each pair; leaflets 7–10 pair, oblong-linear, obtuse, glabrous; spikes axillary, several together, elongated, filiform, anthers tipped with a deciduous gland; legumes cylindric, filled with a mealy pulp. (*W. and A.*)

Hab.—

Parrambo-jammi, Janum-chettu, Chami, Jammi. *Tel.*
Vunni, Parumbay-maram, Vunni-maram? Parumbai-marum, Purrembay-maram. *Tam.*
Shami. *B.*
Sounder. *Mahr.*
Kunda. *Cyn.*
Koduru. *Uriya.*
Kandi, Aghzakar, Se. *N. W. P.*

The late Dr. Stewart remarks that from the stumps of pruned branches and other scars a gum exudes similar to gum Arabic, but he did not know that it is collected or used.

Prunus domestica, *Linn.*

Tree; flowers usually solitary, leaves ovate-lanceolate, convolute; branches unarmed; flowers white. Fruit, variable both in shape and colour.

Var.—ALUCHA. Prunus alucha. *Royle.*

Native names—

Ara-bukhara, Alucha. *H.*

This tree is named by Powell in his Punjab Products as one of the gum yielding trees of the plains.

Prunus puddum, *Roxb.*

Cerasus puddum, *Wall Pl. Asiat. Rar.* pl. 143.

Leaves ovate, lanceolate, acuminated, sharply serrulated; flowers umbellate, umbels on short peduncles, bracteas ciliated, calyx tubular, length of the peduncle, limb of calyx reflected; fruit roundish red.

Hab.—Nepal. Kumaon. Deyra, &c.

Native synonyms—

Puddum, Pudma-ke-gond. *H.*

Enumerated by Powell in his Punjab Products as one of the gum trees of the Hills. It is very probable that the character of the gum would not differ from that of ordinary cherry gum.

Punica granatum, *L.*

W. & A. Prod. i. 327; *Roxb. Fl. Ind.* ii. 499.

Arborescent; leaves oblong, inclining to lanceolate, opposite, calyx 5 cleft; petals five, fruit globose, crowned by the tubular limb of the calyx; seeds numerous, covered with a pellucid pulp; flowers nearly sessile, scarlet.

Hab.—Barbary, cultivated in India.

Pomegranate. *Eng.*
Anar, Gulnar, Darim. *H.*
Dalim, Darmee, Darim. *B.*
Magilan, Madalum. *Tam.*
Dadima-pandoo, Dadima, Puvvu danimma. *Tel.*
Madala. *Mal.*
Delumghedie. *Cey.*
Delema. *My.*
Gangsalan. *Java.*
Ruman, Rana, Kilkul. *A.*
Anar. *P.*

Pomegranate gum appears to have been exhibited at the London Exhibition of 1851 from Vizagapatam.

Sapindus acuminatus, *Wall.*

Royle Illust. p. 139. *Voigt. Cat.* 94.

Tree; leaflets 5–6, oblique lanceolate, acuminate, nearly opposite, shortly petiolate, one shorter terminal; panicle very much branched, terminal.

Hab.—Deyra Dhoon and other Himalayan Valleys. Khassya Mountains. Nepal.

The native names of the tree are—

Dodan, Ritha. *Punj.*

Baden Powell names this as one of the gum yielding trees of the Hills.

Sapindus emarginatus, *W. & A.*

Wi. & Arn. Prod. i. 111; *DCand. Prod.* i. 608; *Wi. Ill.* t. 51.

Tree; petiole pubescent; leaves abruptly pinnate, leaflets 2–3 pairs, oblong, retuse or emarginate, entire; under side very downy; panicles terminal; petals densely hairy on the outside, the margins very woolly, with a small, woolly appendage on each side about the middle; stamens woolly, ovary densely hairy; fruit usually 3-lobed; lobes very hairy on the inside at the insertion of the seeds; flowers small, white. (*Drury.*)

Hab.—Peninsula. Bengal.

Retha, Riti-ka-jhar, Ritha. *H.*
Kankadu, Kunkudu-karra, Kunkudu, Kankadu-chettu, Koonkudu-karra. *Tel.*
Pounanga, Pucha-cotta-maram, Puvandi, Poocheo-cottay, Horingi-maram. *Tam.*
Buro-ritha, Ritha. *B.*
Thalay-marathu. *Can.*
Areeta. *Mahr.*
Rarak. *My.*
Penela. *Cyn.*
Ritha. *Dec.*
Urista. *Sans.*
Mukta-maya. *Uriya.*

One ounce of this gum was sent to the Exhibition of 1862, by Zillah Surgeon Shortt, of Chingleput.

Semecarpus anacardium, *L.*

Wi. & Arn. Prod. i. 108; *DCand. Prod.* ii. 62; *Roxb. Fl. Ind.* ii. 83; *Roxb. Cor. Pl.* t. 12; *Wi. Icon.* 558.

Tree; leaves cuneate obovate, rounded at the apex, whitish beneath; panicles bracteolated; fruit sessile, cordate-ovate, notched on one side under the apex, flowers small, green. (*Drury.*)

Hab.—Coromandel and most parts of India.

Bhela, Bhela-taki, Bhola-taki. *B.*
Bhela, Bhelawan, Bhilama. *H.*
Bhela, Bhola. (*Gualpara.*)
Shayng-cottay maram, Shayrang-cottay. *Tam.*
Bhallatiki? Bhallatamu, Nalla-jidi-chettu, Jidi-chettu, Nellajidi, Jiri, Jidighinzalu, Tummeda-mamidi, Jadi-chetto. *Tel.*
Beladur. *A.*
Chai-bin. *Burm.*
Ghera-mara, Gheru. *Can.*
Kampira. *Mal.*
Arushkara, Bhalataka. *Sans.*
Bhalawan, Bhela. *Dec.*
Bellawa. *Guz.*
Beebwa, Bibooa. *Mahr.*
Bhalleah, Ballia. *Uriya.*

Marking nut gum is described as a coarse black gum, it was sent by Dr. Waring from Travancore to the Madras Exhibition of 1855. The specimen in the collection referred to this source was derived from Salem, and is in amorphous carbonaceous masses with a shining coal-like fracture, and of a dull brownish-black colour. That from Travancore is a brighter black gum, partly soluble, but wholly useless for commercial purposes.

Soymida febrifuga, *Juss.*

Wi. & Arn. i. 122.

Swietina febrifuga. *Roxb. Fl. Ind.* ii. 398. *Roxb. Cor.* 17.

Tree; leaves abruptly pinnate; leaflets opposite, usually 4 pair, oval oblong; racemes axillary, forming a panicle by the abortion of the upper leaves; flowers small, greenish white. (*Drury.*)

Hab.—Central and Southern Provinces. Guzerat.

Rohan, Rohuna, Rohitaka, Rohun. *H.*
Soomi, Sumi, Somida-manu, Somida, Soymeda, Somide-chettu, Somi Somee. *Tel.*
Shem, Choar-kullie-maram, Shem-maram, Sem-maram? Wond-maram, Wonga-maram, Woond-marum, Choar-kalli-marum. *Tam.*
Rohan, Rohuna. *B.*
Rheyn, Rohuni. *Mahr.*
Rohuna, Patranga. *Sans.*
Rouen, Ruhin, *Dec.*
Swamy. *Can.*
Sohu (Sohn?). *Uriya.*

The Choar-kullie gum sent by Dr. Gleghorn to the Madras Exhibition of 1855 is described as a "tolerable" gum.

Spondias mangifera, *P.*

Wi. & Arn. Prod. i. 173; *Roxb. Fl. Ind.* ii. 451.

Poupartia mangifera. *Blume.*

Mangifera pinnata. *Kœnig.*

Spondias amara. *Lam.*

Tree; leaves alternate, unequally pinnated; leaflets 4–5 pair, ovate or elliptic-oblong, abruptly acuminated, oblique at the base, quite entire; panicles axillary and terminal, diffuse; drupe oval, yellow; nut oblong, very hard; flowers small, white. (*Drury.*)

Hab.—Peninsula. Upper India.

Ambalam, Amra. *B.*
Amra, Ambalam. *H.*
Amra, Ambud'ha? Ambodha. *Uriya.*
Kat-ma-aam-maram, Mirri-mangi-maram, Kát-mavu, Kartuma. *Tam.*
Ambara, Amra, Ambála-chettu, Jvuru-mamidi, Pita-vrikshamu, Amatum, Adavi-mamidi, Jadu-mamidi-karra. *Tel.*
Jangli-am, Jungli-am. *Dec.*
Amb. *Mahr.*
Amrataca, Canara-Amra. *Sans.*
Kat-ambalam, Ambalam. *Mal.*
Ambalam. (*Mal.* of Rheede.)
Amra. (*Gualpara.*)
Amseea? *Ass.*
Aembarælla-gaha. *Cyn.*

Of this tree it is said in Lindley's Flora Medica that the trunk when wounded yields large quantities of a mild insipid gum exactly like gum Arabic, and the Rev. J. Long writes in his "Indigenous Plants of Bengal," a gum exudes from the bark in hot weather, it seems very like gum Arabic.

Swietenia chloroxylon, *Roxb.*

Roxb. Fl. Ind. ii. 400; *Roxb. Cor. Pl.* i. 64.
Chloroxylon swietenia, *Wi. & Arn. Prod.* i. 123. *P.*
Tree; leaves abruptly pinnate, leaflets 10–20 pair, pale coloured, unequal sided with pellucid dots; panicle terminal; flowers small, greenish white.
Hab.—Circars, and near Belgaum.

Billoo, Billu-chettu, Billuda, Billu-kurra, Bilugu, Bhalloo-chettu? Billa-karra. *Tel.*
Purush-burus, Vummuray, Mududa, Vum-maai-Porasham, Kodawah-porsh, Kodawah-po-rasham, Vummaiy-maram, Kodawa-pursh, Moodu-dad. *Tam.*
Dhoura. *H.*
Halda. *Mahr.*
Mal-burute, Burutu-gas, Booroochi-gaba. *Cyn.*
Bilugu? Bhayroo, Bheyreé. *Uriya.*
Hulda, Billoo. *Bombay.*
Bilu. (*Kurnool.*)

Satin Wood gum was contributed by Dr. Cleghorn to the Madras Exhibition of 1855. The specimen in the collection from Salem (1873) referred to this source is in irregular fractured and agglutinated tears, very variable in size, brittle, with a shining resinous fracture, translucent, brown, somewhat resembling Babool gum, but more brittle, soluble in water, tasteless or slightly sweet. Mucilage rather turbid, dark mahogany colour, with an odour as of fusel oil. It was a peculiar and remarkable phenomenon which the mucilage of this sample exhibited, in that its surface was in an hour or two covered by a thick pellicle of gum, the upper surface of which became quite dry as if, by rapid evaporation of the water in which it was dissolved, it was returned to the solid state. Although this pellicle was broken up it continued daily to re-form on the surface of the solution.

Another sample in the reference collection is from Ceylon, paler in colour, and in definite rounded shining amber coloured tears.

Beddome says that the tree yields a wood oil. It seems rather inconsistent to hold that a soluble gum and an oleo-resin are alike yielded by the same tree.

Swietenia mahagoni, *L.*

Grieseb. W. I. Fl. p. 131. *Bot. Misc.* i. t. 16, 17.
A high tree, leaflets 6–10, inæquilateral, ovate or ovate-lanceolate, pointed; capsule ovoid.
Hab.—Cultivated in India.

This tree Dr. Birdwood observes yields in Bombay a beautiful silvery gum in great abundance. It is named as one of the contributions of Dr. Cleghorn to the Madras Exhibition of 1855, and is characterised in the catalogue as a "superior" gum. The gum of the mahogany tree is named as one of the products of the island of Jamaica.

Tamarindus indica, *L.*

Wi. & Arn. Prod. i. 285. *DCand. Prod.* ii. 488. *Roxb. Fl. Ind.* iii. 215.
Tree; leaves abruptly pinnated; leaflets numerous, flowers racemose, somewhat orange, sheathed with scarlet, slightly fragrant.
Hab.—Peninsula. Bengal. Assam.

Amli, Imlee, Nuli-ambli. *H.*
Chinta, Chinta-chettu. *Tel.*
Pooli, Poolia-marum, Pulia-maram, Pollium, Polia-maram, Pulli-maram. *Tam.*
Karangee, Kranji, Kamal, Asam, Assam-jawa. *My.*
Seiambala, Sigembela, Maha-si-yambala. *Cyn.*
Tamr-i-hindi? Amblie, Tamar-bindee. *A.*
Tintori, Nuli, Ambli, Amlee, Tentool. *B.*
Ma-gyi. *Burm.*
Huli-shena, Oonara-mara, Hoonsay? Onara-mara. *Can.*
Cheetz. *Mahr.*
Balam-puli, Balam-pollie. *Mal.*
Darakht-tamr-i-hindi. *P.*
Umlika, Tintiree, Tintili. *Sans.*
Tentooli, Koyan. *Uriya.*
Teteli. *Ass.*

Tamarind tree gum was contributed by Dr. Waring to the Madras Exhibition, 1855, from Travancore, and it was also received from Masulipatam. The specimen sent to Vienna in 1873 from Salem is a dirty black gum, mixed with smaller, lighter coloured fragments, partially soluble, containing much impurity.

Tamarix dioica, *Roxb.*

Hook Fl. Ind. 249. *Roxb. Fl. Ind.* ii. 101.
Small undershrub; young branchlets glabrous; leaves sheathing; flower-bearing branchlets about as long as the terminal spikes; flowers diæcious, styles elongated beyond the corolla; flowers very small, rose-coloured.

Hab.—From Sind and Punjab to Assam. The W. Peninsula and Burma, near rivers and on the sea coast.

Asul. *H.*

Mr. Baden Powell thus describes this gum :—"Occurs in nodules, highly friable, "of a granular texture, the nodules appear opaque or a palo yellow, but the little "grains of which the nodules consist are individually transparent, the centre of each "nodule is more transparent and of a red colour; its taste is very peculiar, of a "bitter combined with sweet, like a mixture of liquorice, aloes, and sugar; it is "quite soluble in water." (*Punj. Prod. p.* 397.)

Terminalia arjuna, *W. & A.*

Wi. & Arn. Prod. i. 314.

Pentaptera arjuna. *Roxb. Fl. Ind.* ii. 438.

Tree; bark smooth, branches horizontal, leaves sub-opposite, linear-oblong, with unequally cordate base, smooth, having two sessile glands underneath the base, and not visible when looking at upper surface of the leaf.

Hab.—

Names of the tree are—

Koha, Kowah, Arjun, Urjen, Kahua, Kowah, Cahua, Arjoon, Unjen. *H.*
Eri-maddee. *Tel.*
Vella-marda, Vella-murda, Tella-marda, Kola-murdah, Vellai-marudu. Valley-murdah-maram. *Tam.*
Bili-mutti, Bil-muttee, Belee-waulkee. *Can. H.*
Arjun, Urjoon. *B.*
Touk-kyan, Touk-kyau. *Burm. Tam.*
Urjuna, Urjun-sadra? *Dec.*
Arjoon, Sadura, Azun. *Mahr.*
Arjuna, Kukhooba. *Sans.*
Arjun-sadara. *Bombay.*
Arjuna, Arjoono. *Uriya.*
Orjun. *Ass.*

Arjan-gond Mr. Baden Powell states is a gum found only with druggists. It is of a clear brown colour and quite transparent. A small sample of this gum was sent by Dr. Shortt to the London Exhibition of 1862.

Terminalia belerica, *Roxb.*

Roxb. Fl. Ind. ii. 431. *Roxb. Cor.* 198.

Terminalia punctata. *DCand. Prod.* ii. 13.

Tree; leaves about the extremities of the branchlets, long-petioled, obovate, obtuse or shortly acuminated, quite entire, glabrous; spikes axillary, solitary, almost the length of the leaves; bisexual flowers sessile; male shortly pedicellate, with a large hairy glandular disc in the bottom of the calyx; drupe obovate, obscurely 5-angled, fleshy, covered with greyish silky down; flowers small, greyish green, offensively smelling. (*Drury.*)

Hab.—Peninsula. Goruckpore. Goalpara.

Tani, Tondi-chettu, Tadi-chettu, Katthu-elupæ, Bahadha, Thanddee, Thandi, Bahadrha. *Tel.*
Tani, Tanikai, Tandi-maram, Tani-kaia-maram, Tonda-maram, Cattu-elupa, Vellai-murdoo, Tanikoi, Tandee-marum, Cattoo-eloopœ. *Tam.*
Bherda, Yehela, Taree-mara. *Can.*
Titseim, Phangah, Pan-gan, Ruheera, Pangah, Bankha, Paugah. *Burm.*
Bulu-gaha, Booloo-gass. *Cyn.*
Beleyluj. *A.*
Buhura. *B.*
Bulla. *Dec.*
Beheyra? *H.*
Berda, Yehela. *Mahr.*
Tani. *Mal.*
Beleyleh. *P.*
Bahira. *Sans.*
Bhára, Bahadha. *Uriya.*
Bauri. (*Gualpara.*)

Dr. Birdwood remarks that this tree abounds with gum, forming probably a portion of gum gattie. In the Proceedings Agri. Hort. Soc. India (April 1840), it is remarked that the Bahera tree yields an immense quantity of gum, which appears in quality to resemble the ordinary gum Senegal of commerce and is largely partaken of by the Chooars and Coles as food. It could be collected in large quantities in the Midnapore forests.

Terminalia catappa, *L.*

W. & Arn. Prod. 313.

Leaves about the extremities of the branchlets, short petioled, obovate, cuneate, and attenuated, but at the same time slightly cordate at the base, a little repand, with a large depressed gland beneath on each side of the midrib, near the base; racemes axillary, solitary, simple, shorter than the leaves; drupe oval, compressed, glabrous, with elevated navicular margins, convex on both sides.

Hab.—

Badamie, Budam. *H.*
Budam, Badam. *B.*
Adamarum. *Mal.*
Nattoo-vadamcottay. *Tam.*
Badam-chettu, Vadam, Vodamo-vettilla. *Tel.*
Badamiehindee, Jungli-badam. *Dec.*
Cotumba. *Cey.*
Catappa. *My.*

Country almond gum was contributed from Masulipatam to the Madras Exhibition of 1855. It is known in the West India Islands as Indian Almond Gum. The museum sample is dark coloured, but soluble in great part, containing fragments of bark.

Terminalia tomentosa, *W. & A.*

Pentaptera tomentosa, *Roxb.* ii. 438.
Terminalia crenulata, *W. & A. Prod.*
Pentaptera crenulata, *Roxb. Fl. Ind.* ii. 438.
Pentaptera glabra, *Roxb. Fl. Ind.* ii. 440.
Pentaptera obovata, *DC. Prod.* iii. 14.
Terminalia elliptica, *Willd.*

Tree. Leaves linear oblong, obtuse, somewhat cordate at the base, crenulate, with terminal glands on the midrib; drupes ovoid, coriaceous, winged; flowers small, greenish white.

Hab.—Common in India.

Nalla-maddee, Tella-maddi-chettu, Maddi-chettu, Muddi-chettu, Nalla-maddi, Tella-muddi, Nella-maddi, Tella-maddi, Tella-maddee. *Tel.*
Saj, Asan ? Jungly-karinj, Urjun ? Epne, Jungle-Kameng, Assan. *H.*
Karra-marda, Vel-maroodum-maram, Karai-marudu-maram, Marudam-marum, Currai-murda-maram, Kurai-maroodoo-marum, Marudum-maram, Arre-muti-maram, Karra-maradu, Maradu-maram, Kurroo-mardoo, Kurroo-murdoo, Curry-murda, Kara-Mardah, Carrumaradu. *Tam.*
Kumbuk, Koombook-gass. *Cyn.*
Aeen ? Mairthee ? Keenjul ? Kunjul ? Eyn, Eyne, Ain. *Mahr.*
Took-kyan, Touk-kyau. *Burm.*
Sahajo, Sahaju, Orjoono. *Uriya.*
Kura-marthi-mara, Mairthee, Muddi, Muttee, Ayni, Mutta, Mariee, Muddee, Maroodum, Karrai-mutti, Madi, Matti. *Can.*
Jungli-karing, Piasal, Asan. *Dec.*
Arjuna ? *Sans.*
Aans ? Ashan, Piya-shal. *B.*
Aeen, Maitree. *Bombay.*
Carrimariddy. *My.*
Pannee-moodee ? *Ass.*

The gum from this tree sent to the Punjab Exhibition from Madras is described as a red gum, black outside the pieces. A specimen in this collection sent from Berar in 1873 is in rounded dull brown tears, soft, readily becoming agglutinated, and capable of being sliced with a knife, having a bitter disagreeable taste, and partially soluble in water. Specimen from Mr. Broughton (Madras) is much darker in colour, being pitchy brown, and commercially valueless.

Thespesia populnea, *Cor.*

Wi. & Arn. Prod. i. 54.
Hibiscus populneus. *Roxb. Fl. Ind.* iii. 190. *Wi. Icon.* 8.
Malvariscus populneus, *Gaertn.*

Tree. Leaves roundish cordate, acuminated, entire, smooth; pedicels about as long as the petioles; capsule spherical, depressed, indehiscent; flowers large, yellow, with a purple eye. (*Drury.*)

Hab.—Peninsula.

Native synonyms of the tree are—

Parspipal, Parspippu, Porush, Paris. *H.*
Poresh, Porush. *B.*
Porsung, Pooarasoo. *Tam.*
Ghengheravie. *Tel.*
Parish. *Dec.*
Sooriya-gaha. *Cey.*

The specimen sent from Madras to the Punjab Exhibition is characterised as a red gum. That sent from Coimbatore (No. 2098) to London in 1873 is in irregular elongated tears of a dark pitchy brown colour, shining, of the cherry gum kind, tasteless, but little soluble, swelling in water.

Vachellia farnesiana, *W. & A.*

W. & Arn. Prod. i. 272.
Mimosa Farnesiana, *Roxb. Fl. Ind.* ii. 557.
Mimosa sepiaria, *Roxb. in E. I. C. Mus.*
Mimosa Indicata, *Poir.*

Large shrub or tree, armed with stipulary thorns; calyx 5-toothed, corolla tubular, 5-toothed; stamens distinct, leaves bipinnated, pinnæ 4–8 pair; leaflets linear, 10–20 pair, nearly glabrous, petioles and peduncles more or less pubescent, legumes cylindrical, filled with pulp and two rows of seeds, flowers globular, 2–3 together, each on an axillary peduncle, small, yellow, fragrant.

Hab.—Bengal. Assam. Peninsula.

Kasturi-tumma, Naga-tumma, Kasturi, Oda-sale, Pee-tumma, Kampu-tumma, Arimedamu. *Tel.*
Vedda-valla, Veda-vully-maram, Vaday-valli-maram. *Tam.*
Jalli, Jali-mara. *Can.*
Gooya-babula, Guya-babula. *B.*
Guya-babula. *H.*
Iri-babool. *Mahr.*
Urimeda. *Sans.*
Bavur, Baver. *Sindh.*

It is generally remarked that this tree yields an abundance of gum. Mason says that in the Tenasserim provinces, it produces a gum with all the properties of the gum Arabic of commerce. The Museum sample from Salem is in rounded amber tears, clear, soluble, and resembling a good Arabic gum. A sample from Madras in larger tears is darker in colour and more like Babool gum.

Wrightia antidysenterica, *Br.*

Brown in Wern. Soc. Mem. i. 74.

Leaves obovate-oblong, short acuminated, glabrous, corymbs nearly terminal, tube of corolla six times longer than the calyx, follicles distinct, flowers white, sweet scented.

Hab.—Ceylon. S. India.

La-thou. *Burm.*
Inderjau, Dudhii-ke-lakri. *H.*
Vepali, Veppaula, Veppallay. *Tam.*
Kodoga-pala, Pala-chettu, Giri-mallika, Kalingamu, Kodisa-pala-chettu, Kodisa-chettu, Kodisa-pala, Kola-mukki-chakka, Kutajamu, Manu-pala, Pedda-ankudu-chettu, Palavarenu Ankudu. *Tel.*
Dudkhuri. *B.*
Doodkooru. *Ass.*
Conapola.

The gum of this tree is enumerated amongst those sent from Masulipatam to the Madras Exhibition of 1855.

Wrightia tinctoria, *Br.*

Brown in Wern. Soc. Mem. i. 73. *DC. Prod.* viii. 406.

Nerium tinctorium. *Roxb. Fl. Ind.* ii. 4.

Small tree, leaves elliptic-lanceolate and ovate, obtusely acuminated membranaceous; panicles terminal, lax, many flowered; follicles very long and slender, pendulous; flowers white, fragrant. (*Drury.*)

Hab.—Coromandel. West Coast.

Names of the tree—

Chite-ancaloo, Amkudu, Tedlapala, Tshil-ankaloo. *Tel.*
Pala, Palak, Palavay-raynoo. *Tam.*
Bhoorcooree, Kala-koora, Kalakooda. *Mahr.*

The gum of this tree is said to be used for ordinary purposes. A very small sample was sent from Madras to the Exhibition of 1862.

Zizyphus flexuosa, *Wall.*

Roxb. Flor. Ind. ii. 365.

Large shrub, or small tree, with spiny flexuous branches and unarmed straight branchlets, prickles twin, one very long and straight, the other recurved. Leaves lanceolate obtuse, crenate, smooth; flowers axillary, usually solitary, style deeply bifid.

Hab.—Nepal.

Native names of the tree—

Singli, Simli, Barj, Beri, Ber, Relnu. *Punj.*

The Simli gum, said to be derived from this tree, sent over in 1872 for valuation from Central Provinces was very similar to an inferior Babool gum. It was in irregular masses and broken pieces of a dull dark brown colour, with a lustrous fracture, soluble in water, but forming a dark coloured mucilage very like that of coarse dark Babool gum.

DOUBTFUL GUMS.

Schleichera-trijuga. *Willd.*

Zizyphus-jujuba. *Lam.*

Both the above have been quoted as gum-producing trees. It seems probable that as both of them afford lac, this substance has been inadvertently classed with gums, since no evidence can be found that gum is yielded by either of them. Dr. Birdwood remarks, it is true, that a portion of gum gattic is derived from the *Zizyphus jujuba*, but, on the other hand, the Madras Catalogue of 1855 states that the produce of this tree was not a true gum.

Anona squamosa, *L.*

is also a lac producing tree, which has been credited as yielding a gum, but requires confirmation.

Trichosanthes cucumerina, *L.*

is enumerated in Punjab products amongst gums from Madras.

The doubtful and undetermined gums in the Museum Collection are, a partly soluble shining brown gum in elongated tears from Shemooga referred to *Ficus Indica*.

Dhas gum, referred to a species of *Andersonia*, derived from Madras, is a soluble light coloured gum in rounded tears of the average size of a filbert; it resembles the gum of *Grislea tomentosa*.

Kuda-gond from Ahmedabad is mixed, but in great part consisting of fragments of an amber coloured soluble gum.

Antu-gond from Mysore is a peculiar looking brown gum adhering to pieces of bark, running down in stalactitic masses, partially soluble.

Salhe gum is a partly soluble coarse gum from Chutterpore, mixed with bark and other impurities.

Moorkallee gum from Salem is a rough looking pitchy brown gum, rather friable and of but little promise.

Sect. B.—PSEUDO-GUMS.

Insoluble in cold water, swelling, and forming a pasty mass.

Sub-Sect. a. PALE OR TRAGACANTH KIND.

True Tragacanth gum is not produced in India, its place being supplied by coarser kinds known in some markets as False Tragacanth, and in those of India as *Kutira*.

Cochlospermum gossypium, *DC.*

DC. Prod. i. 527. *Wight in Hook. Bot. Misc.* ii. 357.
W. & A. Prod. 87. *Hook. Kew. Jour. Bot.* vi. 306. *Hook. Fl.* 190.
Bombax gossypium, *Linn.* *Roxb. Fl. Ind.* iii. 169.

A small naked tree, branches tomentose. Leaves 3–8 in. diameter, palmately 3–5 lobed, tomentose beneath; lobes acute, entire; petiole slender. Flowers 4–5 in. diameter, in few flowered terminal panicles, golden yellow, sepals silky. Petals obovate, notched. Capsule 2–3 in., 5 lobed.

Hab.—Dry hills, Garwhal. Bundelkund, Berar. Orissa, and the Dekkan. Also commonly planted near temples.

The following native synonyms for the gum are given in the supplement to the Pharmacopœia of India:—

Kaserâé-hind, Samaghul, qatade-hindi. *Ar.*
Katerâé-hindi. *P.*
Hindi-katera-gond. *H. Duk.*

Although the above gum is generally known in the Indian bazaars by the same names which are applied to the true Tragacanth, yet to render them applicable correctly to the former the prefix Indian or country should be added to them, as is the case above.

By some mistake, the Tragacanth gum is considered in some of the bazaars of Southern India to be the produce of the almond tree and named accordingly in Tamul and Telugu *Badam-pishin* and *Badam-pisunu*. These names are, however, not in much use, being generally superseded by the Hindustani name *Katera*. (*Moodeen Shereef.*)

Of this gum it is stated by Mr. Baden Powell that "it is a semi-transparent white "gum in striated pieces very much twisted and contorted; is a substitute for "Tragacanth, and is used in the trade of shoe making." (*Punj. Prod.* p. 397.)

The specimens in the collection are coarser, darker coloured, and evidently inferior to those of the gum of *Sterculia urens*. They are from Goonah, Chutterpore, and Nagode.

The uses to which this class of gum can be applied in Europe are very limited. They can only replace the commonest and coarsest of Tragacanths, for which there is little demand, or be employed for adulteration at home or abroad, unless some new application could be recommended, either in the dressing of fabrics, or in connexion with some other extended branch of manufacture.

Sterculia urens, *Roxb.*

Roxb. Cor. Pl. i. 25 t. 24. *Roxb. Fl. Ind.* iii. 145. *W. & A. Prod.* i. 63. *Dalz. Bom. Fl.* 23. *Miq. Ind. Bat.* i. p. 2, 172. *Hook. Fl. Ind.* i. 355.
Cavallium urens. *Schott. & Endl. Melet.* 33.

A soft wooded tree with white papery outer bark, inner fibrous; trunk erect. Branches spreading, marked with large scars. Leaves about the ends of the branches 9–12 in. long and broad, rounded, base cordate, lobes entire acuminate; petiole 6–9 in.; panicle dense, terminal ∞ flowered, glandular pubescent; flowers small, yellow, ♀ mixed with many ♂; calyx ¼ in. diameter campanulate 5 parted; lobes acute spreading; filaments 10; follicles 5 oblong radiating, yellow, pubescent, with stinging hairs intermixed; seeds 3–6 oblong, black. (*Hook. Fl. Ind.*)

Hab.—N.W. India. Assam. Behar. Eastern and Western Peninsulas. Ceylon.

The native names are:—

Tabsee, Thabsi, Kavali, Tabasu. *Tel.*
Vellaypootalie, Velle-butallé-maram, Vellay-putallé-maram, Kavali? *Tam.*
Kateria-kuli? Katira, Kur-kutila. *H.*
Buli? *B.*
Kundol. *Mahr.*
Odla. (*Gualpara*), *Ass.*

The gum, says O'Shaughnessy, which exudes spontaneously during the hot season, occurs in large light brown transparent tough masses. Immersed in water they swell like a jelly, but do not dissolve but by protracted boiling. The solution is not adhesive and is destitute of the thickness of solutions of ordinary gum. The uses of this gum are very limited, the want of adhesiveness renders it unsuitable for the arts, while its difficult solubility renders it inferior to most other gums for medicinal purposes. (*Dispens.*)

From time to time samples of this gum have been sent over for valuation, but hitherto no use has been found for it, and consequently it has no appreciable value in the markets. The only purpose for which it has hitherto been considered available is as an adulterant of Tragacanth gum, but hardly as a substitute. Some samples either of this or a very similar gum have appeared in the London and Liverpool markets and sold at a low price as false Tragacanth or Hog Gum. It is doubtful whether it can take any place in British Commerce until, by experiment, some special application has been discovered for it. The uses and trade in genuine Tragacanth gum is exceedingly limited, and for the purposes to which that gum is applied the Kuteera gums do not answer equally well.

On the subject of Kuteera gum Professor Guibourt wrote in 1855 as follows:— This gum is of a very different nature from the preceding (Pseudo-tragacanth), as is evident from its properties, as described in the Hist. des Drogues, t. iii. p. 421. Upon its origin there have been formed many erroneous conjectures, even since the publication of some exact data. Thus, Roxburgh has well said (*Flor. Ind.* t. iii. p. 146) that the gum of *Sterculia urens* had been sent to London as a sort of Tragacanth, but that it was soon found that it could not replace that substance; but he gave no name to this gum. De Candolle was perhaps the first to say (*Phys. Veget.* i. 171) that gum Kuteera is produced by *Sterculia urens*, more recently Endlicher (*Enchiridion*, p. 517) includes amongst gummiferous trees the *Sterculia urens* of India and a *Sterculia Tragacantha* from Africa, which from its name alone must surely be the source of a gum similar to the former. This, moreover, accords with the circumstance that one always finds gum Kuteera mixed in small quantity with gum Senegal. Lastly, Endlicher in the same work (p. 532) informs us that *Cochlospermum gossypium*, a tree which has perhaps been wrongly separated from the Sterculiaceæ, produces a gum called *Kuteera*. From all this it results that several trees either belonging to Sterculiaceæ, or allied to that order, appear capable of producing one and the same gum, which gum is that known as *Kuteera*, and this fact seems to me beyond all doubt, both because the *Kuteera* of Africa is exactly similar to that of India, and also because I observed at the Great Exhibition in London, in 1851, a considerable number of samples of Gum Kuteera, of which some were attributed to *Sterculia urens* others to *S. ramosa*, others again to *Cochlospermum gossypium*. All these gums possessed an odour of acetic acid more or less marked, and showed the remains of a coarsely fibrous bark, but it was the gum of *Sterculia urens*, and *S. ramosa* that was Kuteera of the finest quality. (*Pharm. Journ.* xv. 58.)

As to the value of this gum in the London market, the results of an experiment may be cited. Major Glasfurd, Deputy-Commissioner, Chanda, sent, 11th July 1873, samples of gum of *Sterculia urens* (Kuroo) to England for valuation. It was a good selected sample, as good as could be procured, destitute of all extraneous admixture, and yet of two brokers to whom it was submitted, one reported that its value was 20*s.* per cwt., and the other "no value." A previous sample, from Haidarabad, was submitted to two brokers and two gum merchants with the following reports:—(1.) Low quality of Tragacanth or Hog Gum, partly insoluble, value probably about 30*s.* per cwt., better to separate the pale from the brown coloured pieces; (2.) Tragacanth Hog, large 45*s.* per cwt.; (3.) A gum of nature same as Hoggy Tragacanth, but has never appeared in this market from India, might become saleable, value about 30*s.* per cwt.; (4.) A gum of the Tragacanth order, and very sightly but quite insoluble. If imported this, we think, would be used for adulteration. Of the two samples the former (from Chanda) was cleaner picked, and better than the latter (from Haidarabad), yet the report of the brokers was less favourable. This can only be accounted for on the supposition that low qualities of Tragacanth gum were more in demand in the market at one period than at the other.

The museum collection includes samples from Saharunpore, Ahmedabad, and Bengal, and four specimens from Baroda, named respectively (4033) Kurree gum; (4027) Kada tree gum; (4028) Sawree gum; and (4020) Black Saras gum; all these are similar in character. Two samples of False Tragacanth from Chutterpore are doubtless *Sterculia* gums.

A sample of gum called *Bootalli pisin, Tam.*, and much resembling that of *Sterculia urens* was sent to the Vienna Exhibition from some unknown locality in Southern India. It was described as the gum of *Givotia Rottleriformis*, an Euphorbiaceous tree, as far as we are aware never before credited with furnishing a gum. The specimen was in large stalactitic masses of a dull opaque honey colour, and when placed in cold water swelled greatly but did not dissolve, just as the gum of *Sterculia urens* under like conditions. The native name of *Sterculia urens* in Tamul is given as *Vellaybootali*, so that most probably the name of *Givotia Rottleriformis* is entirely a mistake.

Uvaria tomentosa, *Roxb.*

The first and only indication of this as a gum-producing tree occurs in a communication from Capt. Doveton, Conservator of Forests in the Central Provinces, and bearing date 16th August 1873 (No. 827). This letter accompanied samples of four kinds of gum, of which No. 1 was named "Karee," and said to be produced by *Uvaria tomentosa*. Appended are these remarks:—"Resembling gum of the *Sterculia urens* (or Gooloo), which was valued at a high price by Dr. Forbes Watson in his Memorandum published in Supplement Gazette of 15th February 1873. It has, however, a greater affinity for water. When fully charged with water the gum is

" colourless, a circumstance which, from the remarks in the Memorandum above
" referred to, would render it more valuable than the gum of the *Sterculia urens*,
" which is not destitute of colour. The gum of the Karee appears to be a little less
" tenacious than that of the *Sterculia urens*, though the difference is scarcely
" perceptible."

The question for determination in reference to this gum is not so much its difference from that of *Sterculia urens*, since it is difficult to distinguish the two from each other, but really whether there has not been some error in naming its botanical source, *i.e.*, an Anonaceous tree, hitherto never named as yielding gum of any kind. The native name "Karee" is also suspiciously like that of a *Sterculia*.

Sterculia, *other species*.

Only *Sterculia urens* is usually credited with the production of Kutira gum amongst the numerous species which occur in India. It is very probable that this is not an accurate view, indeed some authors have intimated that Kutira gum is yielded in India by several species of *Sterculia*, of which, in addition to the above, *Sterculia ramosa* has been named, as have also five others, *Sterculia fœtida*, *Sterculia campanulata*, *Sterculia piperifolia*, *Sterculia balangas*, and *Sterculia colorata*. The best interpretation we can put on these names is that *Sterculia piperifolia* is a misprint for *S. populifolia*, and that *S. ramosa* is a doubtful name. The following is the order in which the above occur in Hooker's Flora Indica (pp. 355) :—

I. Sterculia fœtida. *Linn.* Western Peninsula, in the Concan. Malabar. Birma.
II. Sterculia urens. *Roxb.* N. W. India. Assam. Behar. E. and W. Peninsulas.
III. Sterculia balanghas. *Linn.* Through the hotter parts of India.
IV. Sterculia colorata. *Roxb.* Eastern Bengal. Pegu. Western Peninsula.
V. Sterculia populifolia. *Roxb.* Western Peninsula. Coromandel.
VI. Sterculia campanulata. *Wall.* Pegu.

There is no mention of a *Sterculia ramosa*, nor does it occur on any authority but that of Dr. McClelland as occurring in Pegu. It is probably a misprint.

Cycas circinalis, *Linn.*

Roxb. Fl. Ind. iii. 744.

Trunk cylindrical, unbranched, surmounted with a terminal bud, consisting in the male of a cone composed of peltate scales; leaves pinnated, thorny, springing from the apex of the trunk. (*Drury.*)

Hab.—S. Concans. Forests near Trichore.

The native names of the tree are—

Wara-gadu. *Tel.*
Todda-pana. *Mal.*
Sagu. *Moluc. Archipelago.*
Buzur-butoo of Bombay.

Dr. Birdwood remarks that *Cycas circinalis* yields a tragacanth-like gum. This is probably the gummy substance exuding from the stem, said to produce rapid suppuration in malignant ulcers.

Opuntia rubescens, *Salm.*

DCand. Prod. iii. 474.

Plant erect, joints elongated, compressed, reddish green, somewhat tuberculately areolate, unarmed; leaves small, reddish at the base of the white tomentum.

Hab.—Brazil. Cultivated.

This plant, cultivated at the Agri-Horticultural Society's garden at Kirkee, is said by Dr. Birdwood to yield a gum somewhat resembling the false tragacanths in large quantity. As it is only a cultivated plant, its produce is not likely to be widely known or to become of any importance.

Sub-Sect. β.—DARK, OR MORINGA KIND.

In many respects these gums resemble those of the previous section, except that they are darker in colour, often of a vinous-red or brownish purple, or purplish black, and hence not possessed of any commercial value in European markets.

Moringa pterygosperma, *Gærtn.*

W. & Arn. Prod. i. 178.

Guilandina Moringa. *Linn.*

Hyperanthera Moringa. *Vahl. Roxb. Fl. Ind.* ii. 368.

Tree 30–35 feet; leaves 2–3 pinnate, with an odd leaflet; calyx 5 cleft; petals 5, nearly equal, the upper one ascending; filaments hairy at the base; racemes panicled; 5 stamens without anthers; seeds numerous, 3 angled, the angles expanding into wings; flowers white. (*Drury.*)

Hab.—Common in gardens in the Peninsula.

"Most samples of Sohajna," writes Powell, "are of a dark colour, but containing "some pieces of a pale waxen rose pink; the latter pieces are the best. The Jhelam "samples contain beside the pink sort some pieces that are very nearly white." And again, "this gum varies in colour from a red to a semi-opaque pink to almost white."

Moringa gum is one of the products which have borne the name of Mocharras or Mochrus.

One sample of this gum in the Reference Collection is in irregularly rounded pale vinous red tears; it came originally from Madras.

Another sample, believed to be from Travancore, is in longer, more stalactitic pieces, of a brown colour, less decidedly reddish, but of a similar character.

Another sample is much blacker, more earthy, in irregular pieces, with no tint of red, and appears to be either a very bad sample, or of doubtful origin.

The Madras Jury Reports state that this gum is obtained in large quantity, but that it does not dissolve in water, resembling in some respects Gum Tragacanth, for which it might probably be substituted. This is exceedingly doubtful, since the colour would be an obstacle, and it has no other superiority over the Kuteera gums.

Stereospermum suaveolens, *DC.*

DCand. Prod. ix. 311. *Wi. Icon.* t. 1342.

Bignonia suaveolens. *Roxb.*

Tecoma suaveolens. *Don.*

Arboreous, leaves unequally pinnate, 2–4 pairs; leaflets oval acuminate, entire; panicles terminal, loose subbrachiate; calyx 5 toothed, corolla hairy or woolly, capsule siliquiform, cylindrical, septum corky cylindrical; flowers dull purplish, very fragrant; leaves vary from broad ovate, shortly and abruptly acuminate, to oval lanceolate long acuminate, entire, or subserrate, pubescent or glabrous, panicles pilosely viscid or glabrous.

Hab.—Deccan. Bengal. West Coast.

Amongst the gums sent over to the Exhibition of 1862 was a small specimen of this gum from Surgeon Shortt of Chingleput. This is another dull vinous-red gum, resembling that of Moringa. It is in rounded tears mixed with impurities.

Ailanthus excelsa, *Roxb.*

Roxb. Fl. Ind. ii. 454 *Cor. Pl.* i. t. 23. *Wi. & Arn. Prod.* i. 150.

Large tree; leaves abruptly pinnated, tomentose when young, afterwards glabrous; leaflets 10–14 pair, coarsely toothed at the base; petals 5, almost glabrous in the inside; filaments glabrous, shorter than the anthers; calyx 5 cleft, samaræ linear-oblong, 3–5, one seeded; panicles terminal; flowers fascicled, green.

Hab.—Northern Circars. Coimbatore.

The native names for the tree are—

Peda-man, Pedda-man, Pedda-manu, Pey-yápa, Peddu-man, Peddu-manu. *Tel.*

Peru-maram, Peroo-marum, Pethawkoo kurra. *Tam.*

Peru-mara. *Mal.*

Maruk. *Mahr.*

Aralu. *Sans.*

Marook. *Bombay.*

Gormi-kawat. *Uriya.*

A small specimen of this gum was sent to the Exhibition of 1862 by Surgeon J. Shortt of Chingleput, where it was stated to be employed as a remedy in dyspepsia. It is in rather large rounded tears, of a deep vinous-red, externally darker and opaque, only partially soluble. Resembles Moringa gum.

Grevillea robusta, *Cunn.*

DCand. Prod. xiv. p. 381. *Bot. Mag.* 59. pl. 3184.

Hab.—Australia. Sparingly cultivated in India.

In the museum collection is a sample of gum, rather like that of *Moringa* of a vinous-red colour, and but little soluble, said to have been derived from the silk oak tree, *Grevillea robusta*, in Mysore. It has a bright shining resinoid fracture, which it retains; it is much mixed with pieces of friable bark to which it adheres. The tree is Australian, and the gum is only botanically interesting.

Eriodendron anfractuosum, *Wi. & Arn.*

Hook. Fl. Ind. 350. *Wi. & Arn. Prod.* i. 61. *Wi. Icon.* t. 400.

Bombax pentandrum, *Linn.*

Tree 50–60 feet; trunk prickly at the base; branches growing out horizontally from the stem, three from one point; leaflets 5–8 quite entire, or serrulated towards the point, lanceolate, mucronate, glaucous beneath; petals 5, united at the base, filaments joined at the base, each bearing 2–3 versatile anfractuose anthers; style, crowned with a 5–6 cleft stigma; capsule, 5-celled, 5-valved; cells, many seeded; seeds embedded in silky cotton; flowers white, springing from the branches.

Hab.—Peninsula. Travancore.

The native names are:—

Ilavam, Elavum-maram, Elavamaram. *Tam.*
Buruga, Boorooga, Pur, Burugumanu. *Tel.*
Shwet-shimool. *B.*
Hattian, Safed-simal, Katan. *H.*
Shamioula. *Mahr.*
Pania, Paniala. *Mal.*
Imbool, Imbool-gass, Pulim. *Cyn.*

O'Shaughnessy says that the gum is termed "*Huttian ke gond*, and is given in "solution with spices in bowel complaints." Dr. Birdwood remarks that the Gum Kuteera of commerce is the product also of this plant, whilst Baden Powell reports that the sample of *Huttian ke gond* exhibited as such at the Lahore Exhibition was black and untransparent, looking more like a gall than a gum.

Three other samples of vinous-red semi-soluble gums, resembling Moringa gum, are in the reference collection.

SELEMBAI—sent from Madras by Dr. Shortt in 1862, in rounded tears of a deep vinous-red colour.

POONGKALEE—also sent from Madras by Dr. Shortt in 1862, consists of irregular and agglutinated tears of a paler vinous-brown, with a shining fracture.

SHAMGATHALEE—also sent by Dr. Shortt in 1862, is in dirty rounded and irregular tears of a deep vinous-red, and shining fracture.

Macaranga Indica, *Wi.*

Wight Icon. t. 1883.

Tree; leaves stipuled, peltate; stipules paired, broad-ovate, cuspidate; male flowers panicled, glomerate; bracts petioled, glandulose; calyx 3-parted, pubescent; stamens 6–8; female panicles axillary; flowers solitary or paired, pedicelled, bracteate; style 1; ovary 1 celled, capsule covered with resinous points, flowers greenish. (*Drury.*)

Hab.—Neilgherries. Travancore.

Macaranga tomentosa, *Wi.*

Wi. Icon. t. 1949. fig. 1.

Tree, young parts obscurely rusty-tomentose; stipules largish triangular-ovate, acuminate, rusty pubescent, leaves orbiculate-ovate, acute, broadly rounded from a peltate base, rusty tomentose, afterwards puberulous, beneath sprinkled with brown shining glands; inflorescence axillary, half equalling the leaves, straight, shortly branched, racemiform, interruptedly glomerule flowered; bracts denticulate, males rhombovate, many flowered; females afterwards ovate, 1–2 flowered, cuneate at the base; segments of male calyx cuneate obovate; ovary much incurved; stigma subovoid, obtuse, deflexed; filaments 3, free; seeds globose, smooth. (*Drury.*)

Hab.—Malabar. Travancore.

Vuttathamaray. *Tam.*
Putta-thamara. *Mal.*

A gummy substance exudes from the cut branches and base of the petioles of these trees. It is of a light crimson colour, and has been used for taking impressions of leaves, coins, medallions, &c. When the gum is pure and carefully prepared, the impressions are as sharp as those of sulphur without its brittleness. Powdered and made into a paste it is reckoned a good external application for venereal sores. (*Drury.*)

Both trees yield a similar gum.

The gum called *Vutta thamary* reported on in the Madras jury reports is the same substance; this is described as a simple pure gum of a crimson colour, exhibited from Travancore, and used for taking impressions of leaves, coins, medallions, &c. A twig of the tree was sent with the specimens and was identified as *Macaranga Indica.*

The Museum specimen when received for valuation was so much decayed and spoilt by enclosure in an air-tight tin box, that no experiments could be made or its properties tested.

Salmalia Malabarica, *Wi. & Arn.*

Wi. & Arn. Prod. i. 61.

Bombax malabaricum. *DC. Prod.* i. 479.

Bombax heptaphyllum. *Roxb. Fl. Ind.* iii. 167.

Large tree; trunk prickly; leaves palmate; leaflets 5-7 quite entire, acuminated at both ends; fruit oblong; flowers very large, deep red. (*Drury.*)

Hab.—India generally.

The native names are:—

Pulá, Elavum-marum, Pula-maram, Mal-ailas-maram, Mall-elava-marum, Pula-mula-elavu. *Tam.*

Buruga, Konda-buruga, Pinna-buruga, Buruga-manu, Bouroga-chettu? Booraga, Mula-buraka-manu, Buraga, Bor-rooga. *Tel.*

Lepan, La-i. *Burm.*

Katu-imbul-gas, Katoo-imbool-gass, Kattu-imbal, Mahatelam-bu. *Cyn.*

Rakto-shimool, Rakta-shimlu, Rakto-shimal, Rakto-simal. *B.*

Mullelavu, Moollelavoo, Mulu-elavum. *Can.*

Mulelavu, Mulu-elavu, Moul-elavoo. *Mal.*

Rakta-simal, Simbal, Ruckta-sembul. *H.*

Kanta-sair, Sair, Sairi. *Mahr.*

Simbal. *P.*

Simal, Salmali. *Sans.*

Bouro. *Uriya.*

Saur. *Dec.*

So many writers assert that this tree yields a gum, although Dr. Birdwood reports that he was never able to obtain any from the tree in Bombay, that for the present at least it must hold a place amongst one of the probable gum-producing trees of India. The Madras Jury Reports state that it is a pure gum, whilst others assert that it is an astringent exudation, and forms one of the kinds of Mochurras sold in the bazaars.

The sample in the collection from Bombay, 1873, contains some pieces scarcely to be distinguished from the Mochurrus of the *Areca catechu*, mixed with tears of a vinous-red gum, resembling that of some specimens of Moringa gum.

Sect. C.—ASTRINGENT GUMS.

The substances grouped under this section are provisionally so placed, because custom has hitherto sanctioned their recognition as gums, but their properties, and the mode in which they are procured, strengthens the inference that they are wrongly placed. It appears that the same arguments which might be employed against the transference of such substances as Gambir and Catechu to the group of Gums and Resins would apply with almost equal force against the retention of Kino and Mochurrus, and similar astringent extracts or exudations, in that group.

These products are reported upon here under this reservation.

Pterocarpus marsupium, *Roxb.*

Roxb. Cor. Pl. ii. t. 116. *Rox. Fl. Ind.* iii. 234.

A lofty tree with the outer coat of the bark brown, the inner red, fibrous and astringent; leaves sub-bifarious, alternate; leaflets, 5-7 alternate, elliptic, emarginate, above shining and of a deep green colour, from 3 to 5 in. long. Panicles terminal; petals white with a tinge of yellow, long clawed, all waved or crested on the margins; stamens 10, united into one body near the base, but soon splitting into two bodies of 5 each. Ovary, generally 2-celled, legume, long stalked, the under ¾ths orbicular, the upper side straight; the whole surrounded with a waved veined membranous wing, rugose and woody in the centre, generally one sometimes two-celled, seed solitary, kidney shaped. (*Roxb.*)

Hab.—Circar Mountains and forests of Malabar coast.

The native names are—

Dammul akhvaine hindi. *Ar*
Khune sugavashane hindi. *P.*
Hira dokhi, Kholar-manda, Rang-barat. *H.*
Nat-ka-dammul-akhvain. *Duk.*
Kandamiruga-mirattam. *Tam.*
Gandamrugam netturu. *Tel.*
Vennap-pasha. *Mal.*

Pereira remarks on the history of Gum Kino that "in 1757 Dr. Fothergill described " an astringent gum which he supposed (though on very loose evidence) to have been " brought from the river Gambia, and hence he termed it Gummi rubrum astringens " gambiense." In 1774 it was introduced into the Edinburgh Pharmacopœia as Gummi Kino, and in 1787 into the London Pharmacopœia as Resina Kino. It was described under this designation in the third edition of Lewis's Exp. Hist. of the Mat. Med. by Dr. Aikin in 1784. In 1794 Schenck published an inaugural dissertation on it. I have not been able to ascertain why it was called Kino, nor can the precise nature of the substance referred to be now ascertained. (*Per. Mat. Med.* ii. 325.)

East India Kino, sometimes called Amboyna Kino and usually known in the shops as gum kino, occurs in small angular glistening fragments, the larger of which appear almost black, the smaller being reddish. When entire they are opaque, but in thin laminæ are transparent and ruby red. They are brittle between the fingers, soften in the mouth, stick to the teeth, and colour the saliva red. They are inodorous, but have a very astringent taste. Both water and alcohol acquire by digestion on kino a deep red colour. The aqueous decoction becomes turbid on cooling. (*Per. Mat. Med.* ii. 326.)

Kennedy states that the juice is extracted when the tree is in blossom, by making longitudinal incisions in the bark round the trunk of the tree, so as to let the gum ooze down into a receiver formed of a broad leaf so placed and fixed in the bark as to prevent the gum from falling on the ground. From the leaf it is made to run into a receptacle placed under the leaf to receive the gum. When this receptacle is filled it is removed, the gum is dried in the sun until it crumbles, and then filled into wooden boxes for exportation.

East India Kino was analysed by Vanquelin, who found its constituents to be as follows:—

Tannin and peculiar extractive - - -	75
Red gum - - - - - -	24
Insoluble matter - - - - -	1

A. W. Buchner has subsequently shown that catechine is a constituent of kino. In this substance kino owes its power of communicating a green colour to the salts of iron.

Henning finds kino to be composed of the following constituents in the order in which they are here placed:—

Kinoic acid.
Tannic acid.
Pectin.
Ulmic acid.
Inorganic salts with excess of earthy bases.

The salts are principally phosphate of soda, and of magnesia, and carbonate and sulphate of lime. The kinoic acid is of a red colour, more readily soluble in alcohol than in cold water; it gives a yellowish precipitate with perchloride of iron and a brownish one with acetate of lead. Henning further expresses an opinion that catechine is not found in true kino, but only in drugs which, though varieties of catechu, are frequently confounded with the former. (*Per. Mat. Med.* ii. 327.)

Museum specimens are from Madras and Travancore. A black resin from Coorg referred to this source is an altogether different product.

Pterocarpus indicus, *Willd.*

Willd. DC. Prod. ii. 419.
Pterocarpus dalbergioides. *Roxb. Fl. Ind.* iii. 236.
Pterocarpus Wallichii. *W. & Ar. Prod.* i. 267.
Pterocarpus santalinus. *Blanco Fl. Fil.* p. 561 (not Linn.).
Pterocarpus Zollingeri. *Miq. Fl. Ind. Bat.* i. p. 136.

Arborescent; leaflets 7–9 (rarely more) ovate acuminate, racemes subramose, paniculate, stamens monadelphous, at length isadelphous, legume stipitate, oblique, suborbicular, minutely downy, thickened in the middle, veined, unarmed, margin broad attenuated, style lateral.

Hab.—South India and Archipelago. Peninsula. Martaban. Moulmein. Mergui. Malacca. Penang. Andaman Islands. Sumatra. Java. Philippines.

The gum kino of the Tenasserim provinces has long been known, as Rev. J. Mason states. "It was brought to Moulmein by an English merchant from the Shan States, "and stated by him, as our Commissioner at the time informed me, to be the production of the *padouk*, the same tree as the one in Moulmein thus denominated by the "Burmans. Several years before I had directed attention to this tree as producing "an astringent gum resembling gum kino, but the medical officer to whom I submitted "specimens said it was a kind of dragon's blood. However after Dr. Morton came to "the provinces, he tried it in his practice and found it, in medicinal virtues, identical "with the gum kino of the druggists." Mr. Mason next proceeds to the identification of the padouk which he says is a name common to two different species of *Pterocarpus*, but which look so much alike that they are usually regarded as one species. One of these he holds to be the *Pterocarpus Wallichii* of Wight, and the other the *Pterocarpus dalbergioides* of Roxburgh. These two forms are now united by Mr. Bentham under one species. Mason goes on to state that "both these species produce "an astringent gum which has been exported for gum kino, or whether a mixture of "both, which is most probable, I am not able to say, possibly neither. It may be "that *P. marsupium* is found in the Shan States, for it grows in Assam where it would "doubtless be called *padouk* by a Burman. Be that as it may, it is certain that these "provinces can furnish the commercial world with a large quantity of gum kino. "The exudation of our padouk, one of the most abundant forest trees, has been "proved by experiment to possess all the qualities of gum kino, while the product "of the neighbouring provinces, whose only avenue to market is through our territories, has been bought by the London druggists for the gum kino of the Pharmacopœia." (*Mason, Burmah*, p. 485.)
The Museum samples are from S. India, the Tenasserim Provinces, and Rangoon.

Sample of Liquid Kino from this tree, lately received for report from Burma, was submitted to Mr. Daniel Hanbury, and his report thereon is as follows:—The liquid kino from Burma affords when dried a gum kino that is not of good appearance. It is intensely black and opaque, whereas good kino is of a dark reddish-brown, and when in minute fragments is seen to be transparent and of a ruby red.

When dissolved it yields a dusky purplish solution, not as in good kino one of a crimson brown.

This bad colour may be due to *contact with iron* the juice having been collected or evaporated in an iron pot, or sent home in a vessel of tinned iron (which was the case), or it may be occasioned by some change undergone by the juice itself in transit to Europe.

The liquid kino has a rather unpleasant odour, which it does not quite lose even after parting with its water.

It would be better to offer the drug in commerce in the dry state.

The Burmese kino tree, called *Padúk*, and by botanists *Pterocarpus Indicus*, is believed to afford a drug as good as the nearly allied *Pterocarpus marsupium*, or kino tree of Malabar.

The juice should not be boiled down, but dried by solar heat, in vessels of earthenware or wood.

Kino is now (Oct. 29, 1874,) quoted in London at 3*l.* to 5*l.* per cwt.

It should be remembered that it is a drug of but small consumption; and that if large supplies were poured into the market it would greatly fall in price, and very possible become wholly unsaleable.

The Brokers report also on this sample that it is quite unsaleable and valueless in the fluid condition.

Butea frondosa, *Roxb.*

Roxb. Fl. Ind. iii. 244. *Cor. Pl.* i. t. 21. *W. & A. Prod.* i. 261.

Erythrina monosperma. *Linn.*

Middle sized tree; leaves pinnately wifoliate, leaflets large, roundish ovate, rather velvety beneath, corolla papilionaceous; racemes simple, many flowered, lax; calyx segments short, slightly acute, several times shorter than the tube; corolla densely pubescent; vexillum ovate, acute, recurved; keel and alœ incurved; legume flat, thin, with a large solitary seed at the apex; flowers in threes, bright scarlet.

Hab.—Malabar. Circars.

The native synonyms of this gum are given as—

Samaghe-palah. *P.*
Palas-ki-gond. *H.*
Chinya-gond, Kinya-gond, Palas-ka-gond. *Duh.*
Murukkan pishin, Palasha-pishin. *Tam.*
Moduga-banka, Palasha-banka. *Tel.*
Plach-cha-pasha, Murukkin-pasha. *Mal.*
Muttaga-gondu. *Can.*
Palash-gun. *B.*
Palasha niryasam. *Sans.*
Khakar-nu-gun. *Guz.*
Kakracha-gonda, Phalasacha-gonda. *Muh.*
Kaliya melliyam. *Cing.*
Pav-si. *Bur.*

Roxburgh says that "from natural fissures and wounds made in the bark of this "tree during the hot season, there issues a most beautiful red juice, which soon "hardens into a ruby coloured brittle astringent gum."

Dr. Royle, in a communication to the Asiatic Society, observed that the Dhak or Pulas was one of the most common plants, either as a tree or shrub, throughout almost every part of India. Its timber was good, but being usually small only used for fuel, the ligneous fibres of the roots were employed for rope-making; its flowers yielded a dye, and from its bark exuded a powerfully astringent gum, while the lac insect frequently settled on its branches, so that the same labourers might be employed in collecting these various products. Dr. Roxburgh has fully described the red juice which exudes from fissures and incisions in both *Butea frondosa* and *B. superba*, which afterwards hardens into a ruby-coloured, brittle, and very astringent gum, but adds that he was unable to find that the natives made any use of the gum or flowers. Dr. Royle, however, stated that it was well known to and employed by the Indians in N. W. India as a medicine, and was there called Kumr-kus, Dhak-ka-gond, and also Kueni-ka-gond; and he added that from the combination of the astringent with the gummy principle it was particularly eligible in many cases for which astringent medicines were prescribed. It thus probably came to be called emphatically *Kumr-kus*, or binder of the loins. It had been brought to England and submitted to him for examination by Mr. Beckett, who had long resided at Alliggurh in the Northern Doab. Mr. B. hoped that it might prove of sufficient value to become an article of commerce, as it was employed by the natives of North Western India for precipitating their indigo. By Col. Galloway it was further stated that it was likewise employed by them in tanning. Dr. Royle alludes to the fact that Pereira found in one of the old druggist's warehouses in the city labelled as gummi rubrum astringens, which it was remarkable was one of the names by which kino was originally known. Some of this having been sent to M. Guibourt; it was described by him as one of the kinds of kino, by the name Gomme Astringente de Gambie, and which he states as having entirely disappeared from commerce, when it was re-discovered by Pereira as above mentioned. It is very probable therefore that the Butea kino very closely resembled the original kino, the origin of which name M. Guibourt states he has found it impossible to trace; but it is probably not accidental that the name of the Butea kino is Kuenee in N. India, and that its Sanscrit is *Kin-suka.* There can be no doubt, however, that the Butea kino was one of the original substitutes for the African kino, and was sold at a high price. (*Pharm. Journ., V., May* 1846.)

According to Mr. E. Solly this gum when purified by simple solution of water, so as to separate the impurities, consists of—

Tannin - - - - - - -	73·26
Difficultly soluble extractive - - - -	5·05
Gum, with gallic acid, and other soluble substances	21·67

He describes it as being a transparent brittle substance of a brilliant ruby red colour, having a very strong astringent taste, and agreeing with the kino of the shops in most particulars, but differing sufficiently to be easily distinguished from it.

Roxburgh writes of it that "this gum held in the flame of a candle swells, and
" burns away slowly without smell or the least flame into a coal, and then into fine
" light white ashes. Held in the mouth it soon dissolves; its taste is strongly, but
" simply astringent. Heat does not soften it, but rather renders it more brittle.
" Pure water dissolves it perfectly; the solution is of a deep clear red colour. It is
" in a great measure soluble in spirits, but this solution is paler and a little turbid;
" the watery solution also becomes turbid when spirit is added, and the spirituous more
" clear by the addition of water; diluted vitriolic acid renders both solutions turbid
" and caustic; vegetable alkali changes the colour of the watery solution to a clean,
" deep, fiery, blood red. The spirituous it also deepens, but in a less degree." (*Roxb. Fl. Ind.* iii. 245.)

The Museum samples are from Indore, Nagode, Travancore, and Cochin.

Butea superba, *Roxb.*

Fl. Ind. iii. 247. *Roxb. Cor. Pl.* i. t. 22. *W. & A. Prod.* i. 261.

Twining shrub with pinnated 3-foliolate leaves; leaflets roundish, velvety beneath; racemes simple, lax; pedicels about twice the length of the calyx; corolla papileonaceous; legumes flat, compressed, thin, clothed with rusty tomentum, with one solitary seed at the apex; calyx segments shortish, acuminate; vexillum ovate, acute; flowers large, bright scarlet.

Hab.—Travancore Forests. Circar Mountains.

The native synonyms for the astringent gum are the same as those already given under *Butea frondosa.*

There seems to be no appreciable difference between the exudations of this species and *B. frondosa,* both being collected indiscriminately, and probably mixed. A specimen from South India referred to this source is in the Museum Collection.

Butea parviflora, *Roxb.*

Wi. & Arn. Prod. 261. *Roxb. Fl. Ind.* iii. 248.

This tree probably affords a similar substance.

Calyptranthes caryophyllæfolia, *Willd.*

Eugenia jambolana Lam. *Roxb. Fl. Ind.* ii. 484.

Calyptranthes Jambolana. *Willd.*

Syzygium Jambolanum. *DC.*

Syzygium caryophyllifolium. *DCand. Prod.* iii. 270.

Arborescent; leaves oval or oblong, more or less acuminated or obtuse, feather-nerved, coriaceous; cymes panicled, lax, usually lateral on the former year's branches, occasionally axillary or terminal; calyx shortly turbinate, truncated; berry olive-shaped, often oblique; flowers small, white.

Hab.—Peninsula. North India.

Native synonyms—

Jaman, Jamoon. *H.*
Narer, Naredoo, Neradi, Neral, Nerar, Neredu. *Tel.*
Thabyew-tha-byai, Thabyion, Thab-yeh-gjo. *Burm.*
Kalajam. *B.*
Jambool. *Dec.*
Perin-njara. *Mal.*
Nawel, Nawel-marum, Narvell, Jambu. *Tam.*
Jamu, Jamo. *Uriya.*
Jamu, Jamee. *Ass.*
Nœrala-mara. *Can.*
Nerndee. (*Kurnool.*)
Sumra, Katammal, Rulchan. *N.W.P.*

Dr. Birdwood remarks that this tree yields a kino-like exudation. It is quoted previously at p. 11 amongst gums, but its place is uncertain in the absence of more definite information.

Agati grandiflora, *Des.*

W. & A. Prod. i. 215.

Æschynomene grandiflora. *Roxb. Fl. Ind.* iii. 331.

Tree; 30–35 feet; leaves abruptly pinnated, leaflets numerous; calyx campanulate, slightly 2-lipped, corolla papilionaceous; vexillum oval, oblong, keel large, falcate, with petals free at the base and apex; racemes axillary, 2–4 flowered; flowers large, scarlet or white; legumes pendulous, very long, many-seeded, contracted between the seeds.

Hab.—Travancore, &c.

The native names are—

Agati. *Tam.*	Agati. *Mal.*
Agisi, Avisi. *Tel.*	Buka. *Sans.*
Buko. *B.*	

Of this tree Dr. Birdwood remarks that it yields an exudation resembling kino.

Bibliography of Kino.

Pereira, Materia Medica, ii. 325, 329.
Guibourt, Hist. des Drogues, ii. 428.
Stevenson and Churchill, Med. Bot., 3rd ed., vol. v. p. 78.
Pharmaceutical Journal, xiii. p. 77; xii. 497; i. 578; v. 500, 495; iv. 510; i. 578; v. 498; xii. 377; ii. 601; i. 399; xiv. 55.
Royle, Materia Medica, 4th ed. p. 331.
Balfour, Cyclopædia, N° 4725.
Journ. Agri. Hort. Soc. of India, vol. v. p. ii. (May 1846); vol. iii. pt. iv. p. 275.
O'Shaughnessy, Bengal Dispensatory, p. 235, &c.
Flückiger and Hanbury, Pharmacographia, p. 170.

Mucherus.

In this place may be enumerated a substance of some interest, but which remains to this day involved in considerable mystery. It is thus described in the Pharmacopœia of India, "The astringent gummy exudation occurs in opaque, dark brown, knotty "pieces, some presenting a remarkable gall-like appearance, inodorous, of a strongly "astringent taste. It contains a large proportion of tannic and gallic acid." For some time it was asserted to be the gum of *Salmalia Malabarica*, and still the assertion is often maintained but without good and sufficient proof; on the other hand Dr. Birdwood affirms that all his attempts to obtain gum of any kind from *Salmalia* completely failed in Bombay, and he has no hesitation in saying that the Red Cotton tree affords no gum whatever. In his Bombay products the same author states that, "The Mochurrus of Bombay is certainly not the gum of this tree "(*Salmalia*), but a kind of gall produced on the *Areca catechu*. The *Moringa* "*pterygosperma* yields a gum, however, which is exactly like the Mojrus received from "Umritsur;" again he distinctly states that, "the Mochurrus of the bazaars is a gall "from *Areca catechu*." It is undoubtedly true that the irregular, inflated, usually hollow, dark brown excrescences very much resemble galls, but, on the other hand, we can learn of no trace of insect life having at any time been found in the interior, or any orifice of escape detected. Many pounds weight of mochurrus have been broken in search of traces of evidence that the structure is that of a gall, understanding by that term a vegetable abnormal growth produced under local irritation caused by insects of some order or other. They are not galls, but gall-like exudations, for there is every appearance of their having exuded, in a manner similar to gum, in a semi-fluid state, which Dr. Birdwood considers to be the fact. This being admitted both from evidence and probability, all that remains is to determine the tree, which Dr. Birdwood affirms to be *Areca catechu*, and the great astringency of the exudation renders this determination exceedingly probable; whilst Dr. Birdwood affirms that he has himself gathered precisely identical excrescences from the *Areca catechu*. It is probable that the name of Mochurrus may be given to more than one substance, but that the Mochurrus of bazaars, especially in Southern and Western India, is the produce of *Areca catechu*.

Describing a specimen of this substance in his "Punjab Products" Mr. B. Powell writes "this is a specimen of that kind of Mocharas which looks not unlike sembal "gum, but is in reality not a gum at all, but a brown astringent gall blister that "is found on the *Areca catechu* palm. A sample is sent called saigata gond "from Gurgaon. In my own collection there is a sample of this gum which I got at "Sealkot called 'mochras or phul supyari'" (flower of the areca), which last name, though "flower" is incorrect for a gall, yet indicates the origin. This is imported from Hindustan and Bengal, so that mochras has three meanings, 1st, Sohajna gum; 2nd, Sembal gum; 3rd, Areca galls.

Group II.—GUM RESINS.

As already indicated, and as intimated by their name, these substances are separated from the gums on the one hand, and the resins on the other, by partaking of the characters of both.

Sect. A.—EMULSIVE GUM RESINS.

Gamboge to be commercially valuable for all the purposes to which that gum-resin is usually applied must be possessed of the property of mixing and forming an emulsion with water. Although it is true that some of the similar products here grouped with Gamboge are deficient, more or less, in that property, yet it is on account of this deficiency more than any other, that they are preclued from any prospect of attaining a recognised commercial value.

Garcinia mangostana, *L.*

DC. Prod. i. 560; *Roxb. Fl. Ind.* ii. 618; *Hook. Fl. Ind.* i. 260; *Hook. Bot. Mag.* t. 4847.

Leaves elliptic-oblong, acuminate, nerves very numerous horizontal, petiole short, thick; male flowers 1½ in. diam; yellow-red or purple, stamens in four masses, stigma sessile.

A small conical tree 20–30 feet, branches many decussate; leaves 6–10 by 2½–4½ in., very coriaceous, nerves regular, close, marching with an intra-marginal one. Male flowers in 3–9 flowered terminal fascicles, pedicels short; sepals orbicular, concave, persistent; petals, broad-ovate, fleshy; stamens, indefinite; hermaphrodite, flowers 2 in. diam., solitary or geminate at the tips of the young branches; pedicels ⅛ in. thick, woody; sepals and petals as in the male; stamens, many; filaments, slender, connate below; ovary, 4–8 celled; stigma sessile, 8 rayed; ovules solitary; berry, as large as an orange, smooth, dark purple; pericarp, firm, spongy; seeds, large flattened; axil, very fleshy, white juicy. (*Hook.*)

Hab.—Malayan Peninsula. S. Tenasserim. Malay Archipelago. Cultivated in Ceylon and in parts of Madras Presidency.

The native names are :—

Manggusta. *Mal.*
Mangeestun. *Bombay.*
Manggis. *Java, Sunda, Bali.*
Manggos. *Lampung.*
Mangissi. *Bugis.*

The specimen of gum sent to the London Exhibition of 1862 by Mr. A. A. de Wind, from Malacca, as the produce of this tree, somewhat resembled gamboge externally, was in small semi-opaque smooth rounded tears, but would not easily form an emulsion, and could not be used as a pigment.

O'Shaughnessy observes of this species that from the rind of the fruit he had obtained small quantities of fine gamboge.

Garcinia cambogia, *Desr.*

Desvouss in Lamk. Ency. iii. 701. *Hook. Fl. Ind.* i. 261.
Roxb. Cor. iii. 298. *Fl. Ind.* ii. 621. *DC. Prod.* i. 561. *Wi. & Arn. Prod.* i. 561. *Bedd. Fl. Sylv.* t. 85.
Garcinia zeylanica, *Roxb. Fl. Ind.* ii. 621.
Garcinia affinis, *W. & A. Prod.* 101.

Leaves oblong, elliptic, or lanceolate, dark green beneath, nerves very oblique, male pedicels ¼–¾; rudimentary stigma 3–4 lobed, or O.; fruit not grooved to the tip.

A small erect tree, branches drooping. Leaves 2–5 by ½–1½ in., dark green, shining, hardly coriaceous, shortly acuminate, base acute, rarely obtuse; midrib prominent to the middle; veins thick, usually very oblique, ¼ in. apart, reticulate; petiole, ⅓ in. Male flowers, in short axillary fasicles, pedicels thickened towards the tip, often reflexed; sepals with narrow membranous margins, outer ⅙–¼ in. long, inner larger. Petals twice as long as the sepals, thinner, oblong, concave. Stamens, 12–20 or more, adnate to the receptacle into a mass with a prominent centre, free part of filaments $\frac{1}{30}$ in. long. Rudimentary stigmas 3–4, very short, or O. Hermaph. fl. 1–3, terminal and axillary, rather larger than the male, pedicels ⅙–¼ in. Stamens 10–20; filaments unequal, all connate at the base, or in unequal bundles. Stigmatic rays 8–10, tubercled, free nearly to the base, spreading. Fruit the size of a small apple, yellow or red, grooves 6–8, ending about the middle; top flat, depressed, mamilla thick. Seeds 6–8, aril succulent. (*Hook.*)

Hab.—Mountains of the West Peninsula, from Concan to Travancore, Ceylon.

Thwaites states that this species yields (in Ceylon) a yellow insoluble gum, which is hence valueless as a pigment. Beddome recognises as varieties of the tree one found on the Shevagerry Hills, and figured in Wight's Icones, t. 121, and another found in the Conoor and Sisparah jungles, figured by Wight in the same work at plates 960, 961. Mr. Broughton reports (*Drury's Useful Plants*, p. 220), that this gamboge, though produced by a different tree to those which yield the Siam and Ceylon gamboge, appears, nevertheless, exceedingly similar, and to be of fine quality. An estimation of the amount of colouring resin, which is the essential constituent, gave a yield of 76 per cent., the remainder consisting of gum and starch. The specimen I received was in small lumps, and differed thus in external appearance to the commercial specimens I have seen, but in quality it can well compare with them. The price of the Canara gum is 1 rupee per lb.

Two samples are in the reference collection, in external appearance as well as composition very greatly resembling the gum of *Garcinia pictoria*. One of these samples was collected in 1804, amongst the mountains of Wynaad, the other sent more recently from Mysore. It seems very probable that not only do the remarks of Mr. Broughton apply to *Garcinia pictoria*, but that the two specimens just referred to are also derived from that source and *not* from *Garcinia cambogia*, of which Beddome says, "the pigment which exudes from the trunk is semi-transparent, very adhesive, and quite unsuitable as a paint."

Writing in 1847 of the gamboge of the Tenasserim Provinces, the Rev. F. Mason says of this species, "the tree is very common in the Tenasserim Provinces, but the bright yellow exudation it produces is certainly not gamboge;" and, again, he writes, "the exudation of *G. cambogia*, though it will not dissolve in water, dissolves in spirits of turpentine, and forms a very beautiful yellow varnish for tin and other metallic surfaces." It was probably the same substance to which Dr. Helfre alludes when he writes "the gamboge of this country (Tenasserim) dissolves very little with water, and, consequently, does not yield that yellow emulsion as the common Guttifera. It will never serve as a colour, but promises to give a very beautiful varnish."

Garcinia Cowa, *Roxb.*

Fl. Ind. ii. 622. *Hook. Fl. Ind.* i. 262. *DC. Prod.* i. 561. *Wi. & Arn. Prod.* i. 101.

Garcinia Kydia. *Roxb. Fl. Ind.* ii. 623. *Wight Ic.* t. 113.

Garcinia Roxburghii, *Wi. Illus.* 125. *Icon.* t. 104.

Garcinia umbellifera, *Roxb. Wall Cat.* 4864.

Garcinia Wallichii, *Choisy Guttif*, 37.

Garcinia lobulosa, *Wall. Cat.* 4868.

Oxycarpus gangetica, *Ham. in Wern. Trans.* v. 344.

Leaves broad lanceolate, acute at both ends, dark green beneath, male pedicels, $\frac{1}{8}$–$\frac{1}{3}$ in., rudimentary stigma 0.; fruit grooved to the tip.

An erect tree, 60 feet, trunk straight, simple; branches many, slender, lower reaching the ground, bark dark grey. Leaves 3–5 by 1–2 in., veins $\frac{1}{10}$–$\frac{1}{8}$ in. apart, slender, regular, marching with an intra-marginal one. Male flowers in 3–8 flowered rarely axillary umbels, pedicels $\frac{1}{8}$–$\frac{1}{3}$ in. Sepals $\frac{1}{8}$ in. long, broad ovate, thick, fleshy, yellow, with pink on both surfaces. Petals twice as long, oblong stamens many; anthers subsessile, 4 celled. Hermaph. fl. solitary rarely 2–3 axillary, sessile, ovary subglobose, stigmatic rays spreading, papillose, stamens (sterile) in four clusters of 3–8 unequal filaments. Fruit the size of a small orange, dark yellow, 4–8 grooved, and celled, depressed above, tip mamillary. (*Hook.*)

Hab.—Eastern Bengal. Hills near Monghyr. Assam. Eastern Peninsula. Andaman Islands.

This tree is stated, on the authority, we believe, of Dr. H. Cleghorn, to afford a kind of gamboge, differing in colour from ordinary gamboge. There is in the reference collection a curious specimen of a sulphur coloured gamboge run into a small bamboo, so as to form a stick not exceeding an inch in diameter, but its botanical source is unknown; it came from Sarawak, and may possibly be referred to this species; there is also another specimen named *Gutta Kandis*, from Borneo, similar in colour, but consisting of irregular fragments of a larger cake or agglutinated mass of tears. The paleness of colour is a characteristic of this Bornean gamboge. It is rubbed down with more difficulty than good gamboge. There is also in the reference collection a specimen of gum resin resembling gamboge from Java, called *Getah Kawhe*. It is darker in colour than Siam gamboge, and of a redder orange, in irregular lacunose fragments, rather friable, earthy, and not at all shining in fracture, and will not form an emulsion in the same manner as gamboge.

A dull earthy deep orange-coloured specimen from Malacca called Gutta Kandees or gum of the wild Mangosteen, can hardly be derived from the same source as the sulphur coloured gamboge above mentioned, but identification from the specimens is hardly possible.

Garcinia Wightii, *And.*

Anderson in Hook. Fl. Ind. i. 265.

Leaves coriaceous, linear-lanceolate, acuminate; petals equalling the sepals.

Leaves 3-5 by $\frac{1}{2}$-1 in.; veins very slender, oblique, petiole $\frac{1}{4}$-$\frac{1}{3}$ in. Male flower small, sessile, axilly. Sepals equal, thinly coriaceous, orbicular, concave. Petals obovate, very concave, stamens about 20, in a 4-gonal column enclosing a rudimentary 4-gonal style, filaments free above; anthers peltate, dehiscence oblique. Female flowers solitary, sessile. Fruit the size of a small cherry; stigma 4-lobed. (*Hook.*)

Hab.—Forests of southern India.

Of this tree it is stated by Anderson (as above) that the gamboge is very soluble, and yields a good pigment.

Garcinia morella, *Desr.*

Desrouss in Lamk. Ency. iii. 701 t. 405, f. 2. *Hook. Fl. Ind.* i. 264. *DC. Prod.* i. 561. *Hanbury in Linn. Trans.* xxiv. 489, t. 50. *Bedd. Fl. Syl.* t. 86.

Cambogia gutta. *Linn. Fl. Zey.* p. 87.

Hebradendron cambogioides. *Grah. in Bot. Mag.* ii. t. 27.

Garcinia gutta. *Wi. Ill.* i. 126, t. 144.

Garcinia elliptica. *Wall. Wi. Ic.* t. 120.

A middling sized tree, everywhere glabrous; leaves elliptic, with a very obtuse blunt point, and gradually attenuated at the base, about 4 in. long by $1\frac{1}{2}$ to 2 in. broad; petioles about $\frac{1}{4}$ in. long; flowers sessile, several together in the axils of the fallen leaves, calyx of four unequal sepals, the two inner being much larger than the two outer, male stamens about 26, closely packed on a raised receptacle in the centre of the flower without any rudiment of an ovary, anthers on very short thick filaments depressed peltate circumscissile; female flower, stamens 18-20 in one series round the base of the ovary, anthers sterile subquadrate, emarginate at the apex, ovary glabrous 4-celled, crowned with a large sessile 4-lobed stigma, the lobes being 2-3 toothed at the margins; fruit size of a cherry, globose, 4 seeded. (*Beddome.*)

Hab.—South Canara. Ceylon. Tenasserim.

The tree is called—

Gokatoo, Kana-goraka. *Cey.*
Aradal, Punar puli. *Can.*

The gum resin is—

Rubbe-revand, Ausarahe-revand. *A.*
Ghota-ghanba, Gota-ganba. *H.*
Ausarahe-revan. *Duk.*
Makki, Ireval-chinip-pal. *Tam.*
Revalchini-pal. *Tal.*
Revachinni-sira. *Mah.*
Sanatosi, Tanato-asi, Sanatho-asi. *Bur.*
Gokatu, Gokatu-melliyam. *Cing.*

According to Pereira the first notice of gamboge is by Clusius in 1605. He received this gum resin in 1603 from Peter Garet of Amsterdam. It had been brought from China by Admiral van Nech and his companions, and its oriental name was said to be *Ghittaiemou.*

Until very recently the source of the gamboge of commerce was obscure, that of Siam being referred to *Garcinia cochinsinensis,* and that of Ceylon to *Hebradendron cambogioides,* whilst now both are referred to the same species, determined by Mr. Daniel Hanbury to be *Garcinia morella.* Desr. var. *pedicellata.*

"The whole history of gamboge, including the various opinions that have been " entertained as to its source, and the facts that have been made out during two " centuries respecting it, have been admirably held by the learned French pharma- " cologist Professor Guiboart, and a resumé of them given by Dr. Planchon and " Triana." Thus writes Mr. Hanbury, and then he proceeds to state that the gamboge tree having been transported from Siam to Singapore, and desiring to set the matter at rest as to the question of origin of gamboge, he wrote for and obtained from Messrs. D'Almeida of Singapore the necessary materials for determining the species of *Garcinia* cultivated by them, and which is believed to be the real gamboge tree of Siam. In conjunction with Professor Oliver the materials supplied were investigated, resulting in the determination of the Singapore *Garcinia* as a variety (having the male flowers pedicelled) of *Garcinia morella,* which at the same time was believed to be specifically the same as the *Garcinia elliptica* of Wallich. This determination of identity is important in connexion with the remarks of the Rev. F. Mason on the second gamboge tree of Tenasserim. "South of the latitude of the mouth of the "Tavoy," he writes, "and throughout the province of Mergui, there is found on the "low plains at the foot of the hills, and on the banks of the rivers almost down to "tide waters, another species of *Garcinia* that also produces good gamboge. I have "no doubt but it is the tree from which Dr. Griffiths furnished Dr. Wight with "specimens, and which the latter says I refer doubtfully to Wallich's *G. elliptica.* "We will call it then *G. elliptica*, a species which Dr. Wight has on his list of "'species imperfectly known.' The foliation and female flowers are, however, very "well described, and to complete the description I may add the male flowers are "pedunculated, but the peduncles are short, and they might be characterised as sub-"sessile. The anthers, like those of the female flowers, are sessile, depressed, or "flattened above, and delusced circularly. The ripe fruit is globose, and not furrowed. "Neither Wallich, Wight, nor Griffiths appear to have been at all aware that this "species produces gamboge." The gum of this tree he describes "in its appearance "to the eye, and in its properties, as a pigment. I have failed to discover the "slightest difference between it and the gamboge of commerce. It serves equally "well to colour drawings. The Burmese priests often use it to colour their garments, "and the Karens to dye their thread. It is also used by the native doctors in medi-"cine, but I think not extensively." (*Journ. As. Soc.*) Hence it would appear that one of the two gamboge trees of Tenasserim is in reality identical with the gamboge tree of Siam and Ceylon, the other being the *Garcinia pictoria*, Roxb., a closely allied species.

In his observations on Mysore gamboge Dr. Christison cites a letter from Dr. Wight, from which it is evident that Wight knew that the *G. elliptica* produced gamboge. "I have lately received," it states, "flowering specimens of *Garcinia elliptica*, of "which I have published a figure in my Icones as the gamboge tree of Tavoy and "Moulmein, and not improbably of Siam. There was sent along with the specimens "a very small quantity of the juice, which was quite emulsive, although in itself a "very bad and dirty sample. On trial it could be used readily as a pigment, but the "colour was bad, whether naturally so, or owing to dirt with which it was mixed, I "could not make out. The specimen was too small for me to examine thoroughly. "A further and better supply was promised me. But the gentleman who interested "himself in the inquiry was shortly after removed from that coast, and there the "matter dropped. All the other juices of the Guttiferœ I have examined are resi-"nous, and do not mix with water, in short, are not emulsive."

The only account which we possess of the method of obtaining Siam gamboge is that given to König by a Catholic priest residing at Cochin China. According to this statement, when the leaves or branches are broken, a yellow milky juice issues, *guttatim* (hence the origin of the term *Gummi Guttœ* applied to gamboge), and is received either on the leaves of the tree, or in cocoa-nut shells, and from thence is transferred into large flat earthen vessels, where it is allowed to harden during the summer season, and is afterwards enveloped with leaves. The cylindrical or pipe variety receives its form by being run into the joints of the bamboo while it is in the liquid state. A few years since there was an importation of gamboge in the bamboo cylinders (gamboge in the bamboo). Each cylinder or stem was about 21 in. long, and one and a half in diameter, closed at the lower end by the transverse partition of the nodes, and at the upper by a piece of oilskin. In Ceylon, gamboge is obtained by wounding the bark of the tree in various places with a sharp stone when the flowers begin to appear. The cream-like juice which exudes hardens in the sun. According to Mrs. Walker the Cingalese method of collecting it is by cutting pieces of the bark completely off, about the size of the palm of the hand, early in the morning. The gamboge oozes out from the pores of the bark in a semi-liquid state, but soon thickens, and is scraped off by the collectors next morning without injury to the tree, the wounds in the bark readily healing, and becoming fit to undergo the operation again. (*Pereira, Mat. Med.*, 526.)

The following descriptions of gamboge are described by pharmacologists :—

Siam gamboge.—This is the gamboge of the shops. It is brought to this country sometimes direct from Siam, at other times indirectly by way of Singapore, Penang, or Canton. It comes over in boxes, cases, or chests. It presents itself in commerce in three forms. 1st, in rolls or solid cylinders; 2nd, in pipes or hollow cylinders; 3rd, in cakes or amorphous masses. Both the solid and hollow cylinders are known in commerce as pipe gamboge. What is called coarse gamboge consists of the commonest pieces of the above.

α. *Pipe gamboge* consists of cylindrical pieces varying in size from one to three inches in diameter. Some of them are striated from the impression of the bamboo stems, into the hollow of which the gamboge juice has been run, and not unfrequently portions of the stems are still adherent; and on one occasion the gamboge was imported in the stems. The gamboge cylinders are sometimes distinct and covered externally with a dirty greenish yellow dust; at others agglutinated, or even folded, so as to form masses of varying sizes and forms. Pipe gamboge occurs in all qualities, the finest and the worst specimens of gamboge which I ever saw having this form. Fine gamboge is brittle and odourless, it has very little taste at first, but after some time it causes a sensation of acridity in the throat. Its fracture is conchoidal; its fractured surface is opaque, reddish yellow, with a glimmering lustre. It is completely dissolved by the successive action of ether and water. Mixed with a sufficient quantity of water it forms a yellow emulsion, the films of which are excellent microscopic objects for observing the active molecules described by R. Brown. The powder of fine gamboge is bright yellow. The Edinburgh College gives the following characters of pure gamboge:—Fracture somewhat conchoidal, smooth and glistening; a decoction of its powder cooled is not rendered green by tincture of iodine, but merely somewhat tawny. The iodine is employed to prove the absence of starch. Inferior qualities of gamboge are harder, more earthy in fracture; the fractured surface is brownish, or greyish-yellow, frequently with black spots from the presence of foreign bodies which are intermixed. It is not completely dissolved by the successive action of ether and water. Iodine readily detects in the cooled decoction starch by the green colour which it gives rise to.

β. *Lump* or *cake gamboge* occurs in masses of several pounds weight. Its quality is inferior to the fine pipe kind. Internally we observe fragments of wood twigs and air cells. In most of its characters it agrees with the inferior qualities of pipe gamboge, and like this contains starch.

Ceylon or *Cingalese gamboge.*—I am unacquainted with this kind of gamboge, which is unknown in English commerce. Dr. Christison says that as he has seen it it is usually in small irregular fragments, but as originally collected is in flattish round masses, as if moulded in shallow bowls, weighing about a pound or upwards; and it appears to be composed of aggregated irregular tears, with interspaces and cavities which are lines with a dark powdery matter, or with a powder of an earthy appearance. Altogether it seems a very coarse article. It forms with great ease an emulsion nowise inferior in smoothness, and very little, if at all, in liveliness of tint to that of the very best pipe gamboge of Siam. (*Pereira, Mat. Med.* ii. 528.)

Gamboge was analysed in 1808 by Braconnot, in 1813 by John, and in 1836 by Dr. Christison.

—	Siam Gamboge.						Ceylon Gamboge.		
	Pipe.		Cake.		Coarse.				
	First.	Second.	First.	Second.	First.	Second.	First.	Second.	Third.
Resin	74·2	71·6	64·3	65·0	61·4	35·0	68·8	71·6	72·9
Soluble gum	21·8	24·0	20·7	19·7	17·2	14·2	20·7	18·8	19·4
Woody fibre	trace	trace	4·4	6·2	7·8	19·0	6·8	5·7	4·3
Fecula	—	—	6·2	5·0	7·8	22·0	—	—	—
Moisture	4·8	4·8	4·0	4·2	7·2	10·6	4·6	not ascertained.	
	1008	100·4	99·6	100·1	101·4	100·8	100·9	96·0	96·6

Gamboge from Burmah sent to Vienna in 1873 is in solid cylinders, but so soft as to have run together in an amorphous mass. It is much inferior in colour to Siam, and apparently more resinoid. The other museum samples are pipe gamboge from Cambodia, Calcutta, hollow cylinders from Nugger, all of these being imported, as also the specimen of cake gamboge from Ceylon.

In their recent volume Messrs. Flückiger and Hanbury attribute Siam gamboge to the variety *pedicellata* of *Garcinia morella.* Some botanists are disposed to regard this form as a distinct species. From the report of a paper read at the Linnean Society by Dr. Hooker, it would appear that he considers the gamboge of Siam to be produced, in part at least, of the following species.

Garcinia Griffithii, *And.*

Anderson in Hook. Fl. Ind. i. 266.

Leaves very large, ovate, or ovate-cordate, obtuse, nerves distinct.

A tall tree, young branches 4-gonous, pale green. Leaves 12–18 by 4–6½ in. coriaceous, midrib stout, rigid angled, veins very prominent ¾ in. apart, tips free. Male flowers numerous, small, bright yellow, in very short axillary woody branches, pedicels 2 in. Sepals $\frac{1}{10}$ in. long, concave, fleshy. Petals a little larger, oblong, "dull red." Stamens many in a sub-4-gonal sessile flattened mass; anthers sub-sessile, 2-celled. Rudimentary ovary O. Female flowers 1½-celled. Style very short, stigma peltate, slightly convex, obscurely lobed at the margin, rough with elevated papillæ, persistent. Fruit pomiform, faintly ribbed, 2 in. in longest diameter. Seeds ⅔rds pendulous. (*Hook. Fl. Ind.*)

Eastern Peninsula. Mt. Ophir near Malacca.

Anderson says of this that there is in Maingay's herbarium a plant very like it in habit, but described by him as having a circumscissile anther, which is cultivated at Singapore as the true gamboge of Siam. There still appears to be some doubt as to the source of Siam gamboge which Dr. Hooker seems disposed to attribute to this tree.

Garcinia pictoria, *Roxb.*

Roxb. Fl. Ind. ii. 627. *Wight Ic.* t. 102. *Bedd. Fl. Sylv.* t. 87.

A good sized tree everywhere glabrous, leaves elliptic with a blunt rather sudden point at the apex, and gradually attenuated at the base, about 4 in. long by 1¼ broad; flowers sessile, aggregated in the axils of the fallen leaves, male, stamens numerous, closely packed on a fleshy more or less 4-sided receptacle in the centre of the flower, filaments short, anthers depressed, peltate, circumscissile, no rudiment of an ovary; female; staminodia in a ring round the base of the ovary, filaments in 3–4 rarely 5 phalanges, each bearing 2–7 sterile clavate anthers; ovary oblong 4-celled, stigma sessile of 4 verrucose lobes which are 3–4 toothed at the margins; berry size of a large cherry, oval, smooth, oblong, crowned with the permanent stigma. (*Bedd.*)

Hab.—Common in forests of Western India up to 3,500 feet.

The gum resin bears the same native names as already cited under *Garcinia morella* for commercial gamboge, *i.e.*—

Rubbe-revand, Ausarahe-revand. *Ar.*	Reval-chini-pal. *Tel.*
Ghota-ghanba, Gota-ganba. *H.*	Reva-chini-sira. *Mahr.*
Ausarahe-revan. *Duk.*	Sanatosi, Tanato-asi, Sanatho-asi. *Bur.*
Makki, Jreval-chinip-pal. *Tahr.*	Gokatu, Gokatu-melliyam. *Cing.*

The gum resin resembles Siam gamboge in all its physical characters excepting that it occurs in irregular masses of various sizes, in place of cylindrical pieces. It readily forms an emulsion with water, and has been found to contain 77 to 80 per cent. of cambogic acid. (*Ind. Pharm.* 31.)

Some botanists, such as Dr. Anderson, unite this species with *Garcinia morella*, others again, on account of differences in the female flowers, keep them distinct, as Beddome has done. At any rate it will be an advantage to write separately of the gamboge of India, as distinct from that of Siam and Ceylon, and consequently to maintain the name and characters already given of *Garcinia pictoria*, the common Indian form of gamboge tree.

The Rev. F. Mason writes that the hills that bound the valley of the Tavoy river on both sides, from their bases to their summits, abound with a tree which produces a fine gamboge. It is Roxburgh's *Garcinia pictoria* which he knew produced gamboge, but which he said was liable to fade. "As soon as I satisfied myself of the identity of the "trees, by an examination of the inflorescence of our plant compared with Roxburgh's "description, I coloured a piece of paper one band with this gamboge, and another with "the gamboge of commerce, and subsequently exposed both to the weather equally for "more than twelve months, but without being able to discover that one faded any more "than the other." (*Pharm. Journ.* vii. 398.)

The Mysore country seems, however, to be the primary tract of this species, concerning which the following particulars are furnished by Dr. Christison in his "Observations on Mysore Gamboge." While residing not long ago at Cannanore Mrs. Walker learned from a party of the Company's engineers that they had observed what seemed to be a gamboge tree in the jungles of the Coorg country. This party had been for some time employed under the command of Captain Cotton, Lieut. Francis, and Lieut. Walker, my correspondent's son, in tracing a projected road from Mysore, through the unfrequented jungles of the Coorg Ghauts, towards the Malabar coast. In these operations they had to penetrate dense forests abounding with trees producing vast quantities of gums and resins, and, among the rest, with the tree in question. Lieut. Francis states that this tree was seen almost everywhere on the route from Cannanore to the Ghauts of the Coorg country, and that at one place their encampment in the forest was quite surrounded by it. From this gentleman Mrs. Walker received a small specimen of the exudation, which she obligingly forwarded to me, and he added the remark that he could not detect any difference between it and the colour-makers' gamboge in tint, but on putting the brush into the mouth it does not burn as it ought to do.

Dr. Cleghorn also states " when accompanying our superintendent, Captain Le Hardy, " in his annual tour, the gamboge-bearing tree was first observed near the north- " western frontier of Mysore, as we were proceeding through an interesting almost " unexplored country, everywhere affording the most picturesque scenery, covered " with jungle more or less dense, supplying abundant materials for the botanical " traveller, and especially remarkable for the Gairsoopah Falls, where the river is " precipitated over a vertical scarp 900 feet in height. The tree, which grows to an " elevation of 30 or 50 feet, was met with along the tops of the Ghauts from Gairsoopah " to Nuggur, a distance of 50 miles, and I understand it grows in greater abundance " 10 miles southwards of this. It would appear that its range of elevation is between " 2,000 and 3,000 feet, though it has been seen sparingly on the sloping descent for " some distance down the Ghaut. It is generally found in thick jungle overtopped " by the sago palm, and other forest trees of great height. The foliage is dense, the " branches shoot out horizontally, two or three occasionally coming off in the same " plane, and in general contour it reminds me in some measure of the box on a gigantic " scale." In a subsequent letter Dr. Cleghorn adds that he found the tree in still greater abundance as he proceeded southward; and in another he remarks that on leaving the cool and humid range of the Western Ghauts, whose moist and shady slopes seem to afford a congenial climate, he lost sight of the gamboge tree entirely. From a variety of accounts collated by Mrs. Walker, it has evidently an extensive range, having been met with by the engineer party along a tract of hilly country extending nearly 120 miles from north to south. It seems to flower during the rainy season of the monsoon, at the close of which the yellow juice is said to be exuded in greatest abundance. It is called in the Canarese language *Goorgeemara*, or yellow gum tree.

The specimens of the juice sent to Dr. Christison were all in the concrete state. Those sent by Mrs. Walker present on a fresh surface a tawny brownish yellow colour, and glistening waxy lustre, exactly like fine Siam gamboge, and they show a tendency to the conchoidal fracture. They are free of odour and, in correct language, tasteless, but he had not found them destitute of acridity as Lieut. Francis did, for in a few seconds after tasting a small portion an acrid impression was produced in the back of the palate like what is occasioned by Siam gamboge. The Mysore gamboge is emulsive. Mrs. Walker's specimens are not quite so easily reducible to a fine emulsion with water as the superior qualities of that from Siam, but the subsequent specimens of Dr. Cleghorn seemed to him not at all inferior in this respect to the best commercial gamboge. The emulsion presents the peculiar brilliant, intense yellow hue of ordinary gamboge. When it is used as a pigment he was unable to detect the slightest difference in tint from that variety. He put it into the hands of a skilful colorist along with a specimen of the finest pipe gamboge of Siam, of a quality seldom met with in the colour shops, and received an opinion that the Mysore gamboge was not so easily rubbed down, and that upon paper the tint presented the slightest possible shade nearer brown, but nevertheless that it appeared a superior gamboge, undistinguishable from the gamboge of the shops except on the most narrow inspection. In stating this opinion, it is necessary to add that the standard of comparison had obviously been collected with the greatest care, while Mrs. Walker's specimen had not this advantage; analysis is proving it to be not quite pure, and that the specimen from Dr. Cleghorn, which proved to be considerably purer, is evidently more emulsive. (*Christison in Pharm. Journ.*)

Dr. Christison examined the Mysore gamboge and reports that, " it is, like that of " Ceylon, in a great measure soluble in sulphuric ether, to which it communicates a fine " orange colour; and the solution yields, when evaporated, an orange coloured resin " identical in appearance with that from Siam gamboge. From the residuum, rectified " spirit without acquiring any colour removed a substance which was deposited on " evaporation in the form of a dark smoke-brown extract, like that obtained from the " gamboge of Ceylon. The matter unacted on by the spirit immediately formed an " emulsive solution with cold water, which when evaporated left an adhesive transparent

" substance possessing the general properties of gum. Dr. Cleghorn's specimen left " scarcely any residuum after the action of cold water, but that obtained from " Mrs. Walker left about 3 per cent. of particles of fibre and sand. From an analysis " of 20 grains of each sample, a quantity, however, scarcely sufficient to yield very " positive results, the Mysore gamboge seems to have the following composition:—

	I.	II.
" Resin - - - -	77·0	80·0
" Gum - - - -	14·0	14·0
" Colouring matter - -	5·0	5·0
" Fibre and sand - - -	3·0	0·5
" Moisture - - -	not determined.	

" The composition is essentially the same with that of Ceylon gamboge, but indicates " more colouring matter, more resin, and less gum than in the gamboge of commerce." (*Madras Journ. Arts*, ix., 667.)

The Mysore gamboge in the collection is of two kinds, in tears and in cakes.

Gamboge in tears consists of elongated, orange-red tears, which are sometimes agglutinated into irregular shaped masses, and often mixed with impurities. In samples of good quality the tears are of good colour, homogeneous, and clean, breaking with a shining fracture, and easily rubbed down into an emulsion.

Cake gamboge is either in round cakes about an inch in thickness and four inches in diameter, presenting a mottled appearance when broken, the tears of a clearer paler colour being united by a browner resin; or in thinner cakes, of one uniform colour and texture as if formed by evaporation in shallow vessels, of the exuded juice.

Gamboge juice, of a thick creamy consistence, and the colour of yolk of egg, was sent from Mangalore in 1865, and after nine years is still in a similar condition, having been kept in a close corked bottle.

The samples are—cake gamboge from Canara; lumpy gamboge from Wynaad and Madras; gamboge in tears from Madras. (J. Broughton.)

Garcinia pedunculata, *Roxb.*

Roxb. Fl. Ind. ii. p. 25. *Hook. Fl. Ind.* i. 264. *Wi. Ill.* 125. *Icon.* t. 114, 115.

Leaves obovate, obtuse, long petioled.

A tree about 60 feet; bark spongy; leaves 6–12 by 3–5½ in., subcoriaceous; base acute or acuminate; midrib stout; veins ½ in. apart, regular, parallel; tips marched and anastomosing; petiole 1–1½ in.; male flower large pale green, in 8–12 flowered panicles; pedicels 2–3 in. stout, erect; sepals ⅓ in., orbicular, fleshy, inner part narrower; petals narrower, but scarcely longer, oblong; stamens in a 4-angled truncate shortly stiputate mass; anthers 4 gonal, 2 celled; female flower solitary, larger than the male, terminal on a thick 4-angled peduncle ½ in. long; stamens 20–30 filaments in 4 bundles, connate below; ovary globose, stigmatic rays 8–10 spreading; fruit large, yellow, the size of a small melon; seeds 8–10 large, reniform; aril succulent. (*Hook.*)

Hab.—Forests of N. E. Bengal, near Rungpore and Goalpara, and in Silhet, where it is cultivated for its pleasant acid fruit.

In 1862 Mr. V. P. Coelho sent from Mangalore to the London Exhibition a sample of gamboge said to have been derived from this tree, but there is no record of the tree occurring in Western India.

Garcinia Travancorica, *Bedd.*

Beddome Flor. Sylv. t. 173.

Leaves 2½–3½ in., linear oblong, tip rounded, base acute, nerves very slender, horizontal, surface finely recticulated beneath between them, petiole slender, flowers ½ in. diameter.

A large tree abounding in a yellow pigment; branches obtusely 4-angled shining; leaves coriaceous, sometimes broader upwards, dark green above, not shining, pale beneath, with a stout reddish (when dry) midrib and beautifully reticulated surface, margins recurved, petiole ¼–¾ in. Male flowers in three chotomous short few flowered terminal and sub-terminal cymes; sepals orbicular, concave; petals about twice as large, shortly clawed; stamens in four large masses; anthers very numerous, versatile, two celled, cells longitudinally, two valved; rudimentary ovary columnar, with a circular peltate stigma. Female flowers terminal, sub-solitary; penanth of the male; stammodes few, slender, inserted in a hypogynous ring; ovary globose, half concealed by the large convex obscurely lobed stigma; fruit as large as a walnut, sub-globose, contracted into a short thick style with a broad imbricate stigma ⅓ in. diameter; seeds few, large shining. (*Hook. f.*)

Hab.—Western Peninsula, in Forests of Travancore and Tinnevelly.

What is the character of this yellow pigment with which the tree abounds is not stated, but is presumed to be a kind of gamboge, for Beddome remarks, "that every "portion of the tree yields abundance of a bright yellow gamboge which has not been "examined."

Garcinia xanthochymus, *Hook.*

Hook. Fl. Ind. i. 269.

Xanthochymus pictorius, *Roxb. Cor.* ii. 51 t. 196. *Fl. Ind.* ii. 633.

Xanthochymus tinctorius, *DC. Prod.* i. 562. *W. & Ar. Prod.* 102. *Chois. Gutt.* 32.

Leaves large, linear-oblong or oblong lanceolate, acute or acuminate; male flowers fascicled $\frac{1}{2}$ in. diameter shortly pedicelled.

A middling sized tree; trunk straight, branches drooping, angular, leaves 9–18 by 2–4 coriaceous, shining, veins $\frac{1}{4}$–$\frac{1}{2}$ in. apart, reticulated; petiole 1 in. rugose. Male flowers $\frac{3}{4}$ in. diameter in 4–8 flowered fascicles, from the axils of fallen leaves, white; pedicels thickened 1–1$\frac{1}{2}$ in., sepals $\frac{1}{6}$ in. orbicular, concave, fleshy, unequal. Petals $\frac{1}{3}$ in. orbicular, spreading, thin. Stamens in 5 broad bundles of 3–5 alternating with 5 fleshy glands, anthers 2 celled. Hermaphrodite flowers like the male. Ovary ovoid, pointed, usually 5 celled; stigmatic lobes 5, oblong, spreading entire. Fruit the size of an apple, sub-globose, pointed, dark yellow. Seeds 1–4 oblong. (*Hook.*)

Eastern Bengal and the Eastern Himalaya, from Sikkim to the Khasia Mountains and to Burma. Eastern Peninsula. Penang and the Andaman islands. Western Peninsula, the Circars, and from the Bombay Ghats southward.

Roxburgh states that this species yields a large quantity of indifferent gamboge, and Thwaites says that it yields a tenacious gum of no value. At one time Ceylon gamboge was supposed to be the produce of this species, but Thwaites has declared that *Garcinia morella* is the only gamboge yielding species in Ceylon. Beddome observes "the tree yields a gamboge, but I do not know of what quality."

Bibliography of Gamboge.

Drury, Useful Products, p. 222.
Hanbury in Linn. Trans., xxiv. p. 439.
Pharmaceutical Journal, ii. 602, vi. 60, vii. 398, x. 235, 2nd. ser. vi. 349.
Royle's Materia Medica, 4th. Ed. p. 331.
Pereira, Materia Medica, ii. p. 527.
Mason in Journ. Asiat. Soc., Bengal (1847), vii. 661.
White, Voyage to China Seas, p. 250; Boston, 1823.
Companion to Hooker's Botanical Magazine, ii. 233.
United States Dispensatory, pp. 405.
Ainslie, Materia Medica, i. 148. (1826.)
Pharmacopœia of India, pp. 30, 440.
Ure's Dictionary, under "Gamboge."
Planchon & Triana, Memoire sur la Famille des Guttiferes in Annales des Sci. Nat., 4th ser. xiii. p. 306, xiv. 226, xv. 240, xvi. 263.
Graves' Hortus Medicus, 36.
Guibourt, Hist. Nat. des Drogues, iii. 556.
Balfour's Cyclopædia, under "Gamboge."
Madras Journal of Literature and Science, 1839, p. 123.
Lindley's Flora Medica, p. 113. (1838.)
O'Shaughnessy's Bengal Dispensatory, p. 235, &c.
Flückiger and Hanbury, Pharmacographia, p. 77.

SECT. B.—FŒTID GUM RESINS.

These are, for the most part, the produce of certain umbelliferous plants, exuding in a milky juice from the wounded root, and drying by exposure to the atmosphere. They have a more or less fœtid alliaceous odour, which culminates in Asafœtida, and is just recognisable in Ammoniacum passing through intermediate stages in Sagapenum, Galbanum, and Opoponax. The history of some of them is obscure, and the locality of production doubtful, but none of them are veritable Indian products, being found only in Indian commerce as imports from other places.

Narthex Asafœtida, *Falc.*

Ferula Asafœtida, *Linn.* *Bot. Mag.* t. 5168; *Balf. Trans. Roy. Soc. Ed.* xxii. t. 20, 21.

Root perennial, tapering, ponderous, increasing to the size of a man's arm or leg, covered with a blackish coloured bark, beset near the top with many strong rigid fibres, its internal substance white, fleshy, abounding with a thick milky juice, which has an excessively strong fetid, alliaceous smell. Stem 2 or 3 yards high or more, 6–7 in. in circumference at the base, smooth. Radical leaves nearly 2 feet long. Fruit flat, thin, reddish brown, like that of the parsnip, only rather larger and darker. (*Kœmpfer.*)

Hab.—Extreme Persia. Turkistan. Affghanistan, and the extreme Western Himalaya.

The following native synonyms are given by Moodeen Sheriff:—

Hiltit. *Ar.*
Hing. *Mah., H., B.*
Hing, Vagarni. *Guz.*
Angozah, Angusht-gandah. *P.*
Hing. *Duc.*
Kayam, Perungayam. *Tam.*
Inguva. *Tel.*
Perun-gayam, Kayam. *Mal.*
Perunkayam. *Cing.*
Hinguhu, Ramatham. *Sans.*
Shinkhu, Shingu. *Burm.*

The gum resin occurs in irregular masses, partly composed of tears moist or dry. The colour of a freshly cut or broken piece is opaque white, but it gradually becomes purplish-pink, and ultimately dull yellowish or pinkish brown. Taste bitter acrid; odour fœtid alliaceous and persistent. It dissolves almost entirely in rectified spirit. (*Ind. Pharm.* p. 102.)

Another form of Asafœtida is not uncommon in India, which consists of a ferruginous dry powder intermixed with small fragments not much exceeding the size of a pea, and altogether of a purplish ferruginous colour not unlike in appearance ferruginous clay. There is the characteristic persistent alliaceous odour, but less strong than in other kinds.

Dr. Royle observes in his illustrations that "having obtained in the bazaars two "species of seed as those of the Asafœtida plant, and seeing others in Dr. Lindley's "possession from Beloochistan, it is probable that more than one species of *Ferula* "yields this gum resin in the different parts of Persia, Caubul, and Bokhara." (p. 231.)

Burns in his travels into Bokhara states that the young plant is eaten with relish by the people, and that sheep crop it greedily. The oldest plants are most productive, and those under four years old are not considered worth cutting. At the season when the leaves begin to fade the earth is removed from about the top of the root, and the leaves and stem, being twisted off near the base, are thrown with other vegetable matters over the root in order to protect it from the sun. After some time the summit of the root is cut off transversely, and the juice which exudes having been scraped off another thin slice is removed in order to obtain a fresh surface for exudation. This process is repeated at intervals till the root ceases to afford juice, and perishes. During the whole period of collection, which occupies nearly six weeks, the solar heat is as much as possible excluded. The juice collected from numerous plants is put together and allowed to harden in the sun. (*U.S. Dispens.* 146.)

Kœmpfer divides the business of collecting into four parts; the first begins about the middle of April, and consists of digging the earth about the root, removing the leaves and fibres, which are afterwards laid over the root to defend it from the sun. The second commences on the 25th May. Each collector is provided with a sharp knife to cut the root, a broad iron spatula to scrape off the juice, a cup fixed to his thigh to receive it, and two baskets hung over his shoulders upon a pole. The top of the root is then cut off transversely, and on the third day (*i.e.*, the 27th May) the juice is scraped off and put into the cups. A fresh incision is then made, and the juice removed the day but one following (*i.e.*, the 29th May), when they again cut the roots. The cups are from time to time emptied into large vessels. The juice is exposed to the sun to become harder, and is conveyed home in the baskets. The third and fourth acts are mere repetitions of the second. The third commences about the 10th June, the fourth about the 3rd July. Except after the last operation, the roots are carefully defended from the sun, after each incision by covering them with leaves. (*Pereira, Mat. Med.* 175.)

M. Buhse, as quoted in Royle's Materia Medica, states that the mode of collecting Asafœtida, in Persia, at the present day, is exactly the same as that described by Kœmpfer, 160 years ago.

The plant would appear to occur, as Dr. Falconer writes, in the greatest abundance in the provinces of Khorassan and Laar, in Persia, and thence to extend on the one hand into the plains of Toorkestan, on the Oxus, north of the Hindoo Koosh mountains, where it seems to have been met with by Sir Alex. Burns, and on the other to stretch across from Beloochistan, through Candahar, and other provinces of Affghanistan, to the eastern side of the valley of the Indus, where it stops in Astore, and does not occur in great abundance. The whole of this region, which constitutes the head quarters of the gum-bearing Umbelliferæ, possesses the common character of an excessively dry climate, indicated in Berghaus's hygrometric map, in Johnston's Physical Atlas, by a belt of white. To this Dr. Royle adds that Asafœtida is now cultivated in Affghanistan and the Punjab. It is conveyed on camels into India, across the Punjab and the Bhawulpore, and is sold in large quantities at the Hudwar fair. It is also conveyed down the Indus, and by the Persian gulf to Bombay. (*Mat. Med.*)

When figuring the *Narthex Asafœtida*, in the Botanical Magazine, Hooker observes that "the plant, here figured for the first time from perfect specimens, is one of the several now known to yield the well-known fœtid gum resin Asafœtida, though whether it be, as Falconer supposed, *A. Disgunensis*, indicated by Kœmpfer, and figured in his Amœnitates Exoticæ (p. 535) is still a disputed point. That it yields excellent asafœtida in the form of a copious milky juice, which is collected and exported to Europe in great abundance is clearly made out by Dr. Falconer, who discovered it in Western Tibet, north of Kashmire, in 1838, and sent seeds to the Royal Botanic Gardens of Edinburgh in 1839, where the plant flowered and ripened its fruit last year. It would be impossible to discuss here the vexed question of the history of the origin of all the asafœtidas, nor would the discussion be very profitable. It is certain that Kœmpfer had two plants (species or varieties) in view, from different countries, that his descriptions and drawings and specimens (in the British Museum) do not tally, and that though Dr. Falconer considers his plant one of Kœmpfer's, other botanists do not. Just now, too, we have received at the Museum of the Royal Gardens superb specimens of a very different gigantic Umbellifer, from the Imperial Academy of St. Petersburg, as a true Asafœtida of commerce; it was collected by M. Borsczhow in sandy places on the Steppes east of the Caspian, where it attains a height of nine feet, and yields abundance of excellent asafœtida. Professor Bunge has called this plant *Scorodosma fœtidum* (characterised generally by the absence of vittœ), and M. Borsczhow, who recently visited this country, informs us he believes it to be the Khorassan plant figured by Kœmpfer, and of which fruits are in the British Museum. The same gentleman kindly informs us further, that he considers the Thibetan plant to be quite distinct (in which we entirely concur), and that the *Scorodosma* is probably also found in Eastern Persia. (*Bot. Mag.* 5168.)

Here, then, we have two distinct recognised sources for Asafœtida, one the *Narthex Asafœtida*, as determined by Dr. Falconer, which might be called the "Tibetan Asafœtida plant," and the other the *Scorodosma fœtidum*, discovered by M. Borsczhow, which might be named the "Caspian Asafœtida plant." There is not the slightest doubt that both of these plants yield good Asafœtida, and that the produce finds its way into the markets, but probably that of the former and not of the latter enters into the commerce of India.

As to its occurrence in N. India, the late Dr. Stewart found the plant at Khágán (Jhelam basin) at about 6,000 feet, and Cleghorn mentions specimens of it as being brought to him on the Upper Chenab at over 8,000 feet. It is also given by Aitcheson as growing in Lahoul, 10,000 feet, not much further up the Chenab, but I find from Mr. Jaeschke that this was a *Dorema*. Cleghorn also states that Dr. Falconer sent seeds of the plant from Iskardo to Mussooree and England. Dr. Adams states that he saw the plant in Kashmir, and that loads of it are taken to Sirinuggar; but Dr. Elmslie assures me that the plant is not known there. And it is hardly likely to be common or its product known as derived from it, for what Asafœtida is used in the Punjab Himalaya is imported from Affghanistan, viâ the plains. In Affghanistan the plant grows wild abundantly on the plains to the west of Kilati-Ghilzi (7,000 feet), and the chief supply is obtained from the hills to the north of the Bolan pass, and about the Halmand (perhaps 3,500 feet), according to Bellew. He states that in April and May, incisions (chiefly long cuts which produce lumps, sometimes pricks which yield tears) are made across or round the crown of the root. Gum exudes and is collected for a fortnight from these, to the amount of from one ounce to two pounds (?) weight from a single plant. Bellew also mentions that in Affghanistan the leaves, which likewise have the alliaceous smell, are used as a vegetable, and the succulent part of the young stem is eaten, after roasting, with salt and butter. The Affghans only use the gum medicinally, but very large quantities of it are annually imported into India to be used as a condiment by the natives. Davies estimates the yearly import, viâ Peshawur, at 200 maunds. (*Punjab Plants*, p. 106.)

Some confirmation of the process of collecting Asafœtida will be found in the following quotation from report by Dr. Bellew, formerly attached to Mission to Kandahar, and which is included as Appendix vii. to Davies' Report. "The frail vaginated stem, "or the lower cluster of sheathing leaves, the former belonging to old plants, and the "latter to young ones, is removed at its junction with the root, around which is dug a "small trench about six inches wide and as many deep. Three or four incisions are "then made round the head of the rod, and fresh ones are repeated at intervals of three "or four days; the sap continuing to exude for a week or fortnight, according to the "calibre of the root. In all cases, as soon as the incisions are made, the root head is "covered over with a thick bundle of dried herbs or loose stones as a protection against "the sun; where this is not done, the root withers in the first day, and little or no "juice exudes. The quantity of asafœtida obtained from each root varies from a few "ounces to a couple of pounds weight, according to the size of the roots, some being "no bigger than a carrot, whilst others attain the thickness of a man's leg. The "quality of the gum differs much, and it is always adulterated on the spot by the col-"lectors before it enters the market. The extent of adulteration varies from one-fifth "to one-third, and wheat or barley flour, or powdered gypsum, are the usual adulte-"rants. The best sort, however, which is obtained solely from the node or leaf bud "in the centre of the root head of the newly sprouting plant, is never adulterated, and "sells at a much higher price than the other kinds. The price of the pure drug at "Kandahar varies from four to seven Indian rupees per man-i-tabriz (about 3 lbs.), "and of the inferior kinds from one and a half to three and a half Indian rupees per "'man.' The asafœtida is commonly used by the Mahommedan population of India "as a condiment in several of their dishes, and especially mixed with 'dal.' It is not "an article of general consumption in Affghanistan, though often prescribed as a "warm remedy for cold diseases by the native physicians, who also use it as a vermi-"fuge. The fresh leaves of the plant, which have the same peculiar stench as its "secretion, when cooked are commonly used as an article of diet by those near whose "abode it grows. And the white inner part of the stem of the full grown plant, "which reaches the stature of a man, is considered a delicacy when roasted, and "flavoured with salt and butter."

Subsequently to this, and in another communication, the same authority states that "there are two sorts of asafœtida; that which occurs in tears is the gum that exudes "drop by drop from incisions made in the top of the root. The lump asafœtida is that "which exudes when the root is sliced across, and the juice coagulates." He adds also that there are several umbelliferous plants, especially on the slopes of the Safaid Koh, which exude a milky juice, but which is not collected. Mr. Powell states that the lump sort is the commonest in the bazaars. He had found thin slices of the root mixed with samples of the gum. Asafœtida was brought to him of two colours, one a dirty pale brown, the other with a red or salmon colour, but both similar in texture.

There is a variety of asafœtida called "stony asafœtida," on account of its containing fragments of gypsum, about 50 per cent.; this may be caused by the gum being allowed to drop on the ground of the hill side, where gypsum occurs, and then the whole scraped up together—gum, earth, and all. In the Great Exhibition of 1851, there was an asafœtida, unlike others, of a brown coloured pellucid appearance, full of bits of stalks. This renders the supposition probable that asafœtida is produced by more species than one.

The "Makhzan-ul-adwiya" enumerates two kinds, one called *tib* and one *munattan. Tib* is white asafœtida, called also *kolahpar. Munattan* is dark asafœtida, called in Persian *kamat;* the former is the best kind. In India asafœtida is chiefly obtained overland by the Bolan pass, and by the Peshawur passes, but large quantities are shipped from various parts of the Persian Gulf, and are thence taken to Bombay for export to Europe; the smell is so offensive that sometimes ships are exclusively employed in the Persian gulf to carry this gum. (*Punj. Prod.* p. 406.)

MacCulloch says that commercial asafœtida is imported packed in mats, casks, and cases, the latter being in general the best. It should be chosen clean, fresh, strong, scented, and of a pale reddish colour variegated into a number of fine white tears, which when broken should somewhat resemble marble in appearance, and after being exposed to the air should turn a violet red colour. That which is soft, black, and foul should be rejected. It is chiefly re-exported, being used in England only as a medicine, but in France also in cookery. In London its price varies from 12*s.* to 4*l.* per cwt.

Dr. Pereira has given a summary of the History of Asafœtida which may be cited here with advantage: He says, "it is uncertain at what period asafœtida was first known " or described. The difficulty in determining its history arises from the confusion " which has existed with respect to the *succus cyreniacus* and *asafœtida*. By many " writers the two substances were considered to be identical, but this opinion seems " now to have been satisfactorily disproved by the discovery of the plant called by " the Greeks *σιλφιων*, by the Romans *laserpitum*, which yields the cyrenaic juice, and " which agrees tolerably well with the rude figures struck on the cyrenean coins. " It would appear, however, that the cyrenaic juice becoming scarce, the ancients " employed some other substance of similar though inferior properties as a substitute, " and to both of these they applied the term laser. For many years, says Pliny, this " plant (laserpitium or silphion) has not been found in Cyreniaca, because the " publicans (or farmers of the taxes) who rent the pastures, finding it more profitable, " destroy it as food for cattle. One stalk only, found in our days, was sent to the " Emperor Nero. We may know when cattle meet with the young shoots of it, by " the sleeping of the sheep when they have eaten it, and the sneezing of the goats. " For a long time past the only *laser* brought to us is that which is produced " abundantly in Persia, Media, and Armenia, but it is far inferior to the cyrenaic." It is not at all improbable that the *laser* of Persia may have been our asafœtida. The word *asafœtida*, says Murray, seems to have been introduced by the monks into the school of Salernum. But it appears to have been of oriental origin, and may be, as some have suspected, derived from the word *laser*. Nicolaus Myrepsus, almost the last of the Greek physicians, and who lived, according to Sprengel, about 1227 A.D., speaks of *ασαφιτιδα*. There are two kinds of *Assa*, says Avicenna, one fetid the other odoriferous. (*Mat. Med.* ii. 174.)

The odour of asafœtida is alliaceous, extremely fetid, and tenacious, the taste bitter, acrid, and durable. The effect of time and exposure is to render it more hard and brittle, and to diminish the intensity of its smell and taste, particularly the former. Kœmpfer assures us that one drachm of the fresh juice diffuses a more powerful odour through a close room than one hundred pounds of the drug as usually kept in the stores. Asafœtida softens by heat without melting, and is of difficult pulverization. Its specific gravity is 1·327. It is inflammable, burning with a clear lively flame. It yields all its virtues to alcohol and forms a clear tincture, which becomes milky on the addition of water. Macerated in water it produces a turbid red solution, and triturated with that fluid gives a white or pink coloured milky emulsion of considerable permanence. In 100 parts Pelletier found 65 parts of resin, 19·44 of gum 11·66 of bassorin, 3·60 of volatile oil with traces of supermalate of lime. Brandes' obtained 4·6 parts of volatile oil, 47·25 of a bitter resin soluble in ether, 1·6 of a tasteless resin insoluble in ether, 1·0 of extractive, 19·4 of gum containing traces of potassa and lime united with sulphuric, phosphoric, acetic, and malic acids, 6·4 of bassorin, 6·2 of sulphate of lime, 3·5 of carbonate of lime, 0·4 of oxide of iron and alumina, 0·4 of malate of lime with resin, 6·0 of water, and 4·6 of impurities consisting chiefly of sand and woody fibre. The odour of the gum resin depends on the volatile oil, which may be procured by distillation with water or alcohol. It is lighter than water, colourless when first distilled, but becoming yellow with age, of an exceedingly offensive odour, and of a taste, at first, flat but afterwards bitter and acrid. It contains, according to Stenhouse, from 15·75 to 23 per cent. of sulphur. Hlasiwetz considers it as a mixture in variable proportions of the sulphuret and bisulphuret of a compound radical, consisting of carbon and hydrogen. (*U. S. Dispens.* p. 147.)

Dr. Ure states as one result of his examinations of asafœtida that a specimen in tears had a spec. gr. so low as 1·25, and it left after incineration in a platinum capsule only 5 per cent. of earthy and alkaline matter, chiefly phosphate of lime with a little carbonate of potassa. Good common asafœtida, milky white, semi-hard, has a spec. grav. of 1·31 and affords about 12 per cent. of incombustible ashes consisting of silica, carbonate, and phosphate of lime, with a trace of potassa. Ordinary massive asafœtida has a spec. grav. of 1·35. It affords 25 per cent. of incombustible ashes consisting in a great measure of siliceous and calcareous gravel from the soil round the root of the plants which yielded the asafœtida. (*Pharm. Journ.* i. 464.)

The specimens of this gum resin in the museum collection are from Calcutta, Bombay, and Cabul.

Ferula Persica, *Willd.* (?)

Willd. sp. Pl. i. 1413. *Bot. Mag.* t. 2096. *Steph. and Church.* t. 199.

Hardy plant with perennial root and annual stems. Rises with an erect, round, smooth, slightly striated stem, having only one pair of imperfect leaves about the middle, to the height of three feet. Branches naked and spreading, the three lower ones alternate, supported by the concave membranaceous petiole of the imperfect leaf, the upper ones in whorls, all of them supporting a many rayed sessile, plano-convex terminal umbel, besides from 3–6 branchlets placed on the outside bearing compound umbels. Stem, surrounded at the base with six radical leaves, somewhat glaucous, petioled, spreading, three lobed, ovate, many times pinnate, with incised acute decurrent leaflets of a deep green colour. Footstalks of the lower leaves flat above, with a raised line running longitudinally through the middle of them. Umbels 20–30 general spreading rays, and 10–20 partial ones, with subsessile florets. Florets of the sessile umbels fertile, of the peduncled umbels mostly abortive. Petals yellow, equal, flat, ovate, at first spreading but afterwards reflexed with the tip ascending; filaments awl-shaped, longer than the corolla, curved inwards, and bearing roundish anthers. Germen inferior turbinate, with two reflexed styles and an obtuse stigma. Fruit oval, compressed, marked with three distinct ribs. (*Steph. and Church. Med. Bot.*)

Stephenson and Churchill state that about sixty years ago seeds of *Ferula Persica* were sent to Pallas from the mountains of Ghilan in Persia supposed to be those of the plant producing asafœtida. From these several plants were raised by the Professor at Saint Petersburg, and two of them were sent by Dr. Guthrie to the late Dr. John Hope, Professor of botany in the University of Edinburgh. Both these roots were planted in the open air in the Botanic Gardens there, but one of them died; the other did well and produced seed, and from this source, as Dr. Sims rightly conjectures, spring the plant in the garden of the Apothecaries Company at Chelsea, from which the figure in the Botanical Magazine (t. 2096) and that which embellishes our work (t. 199) was made. Though this is not the plant described by Kœmpfer as yielding the officinal asafœtida, it seems probable that there are various species which afford a gum to which the name of asafœtida has been applied, but that the plant mentioned by this author is the one in which it exists in the greatest degree, or at least which is usually employed for medicinal purposes. That the Ferula Persica produces asafœtida seems confirmed by the strong smell of that drug which pervades the whole herb; and Dr. Sims informs us that he has collected small globules of true asafœtida that had exuded from the stem. The gum resin known in the shops under the name of Sagapenum, which is brought to this country from Alexandria, is supposed by Willdenow to be the produce of *Ferula Persica*. It is, indeed, mentioned by Dioscorides as the concrete juice of a species of Ferula, but of what particular species without authentic materials it is impossible for botanists to decide. (*Med. Bot.* iv.)

Royle (*Mat. Med.*) says that there is no doubt that the seed of this plant has been sent from the North West of Persia as that of the asafœtida plant; but there is no proof, nor indeed is it probable, that it yields any of the asafœtida of commerce. The gum resins of these Umbelliferæ are too similar to each other for any but experienced pharmacologists to determine between inferior asafœtida and varieties of Sagapenum or other gum resins.

On the other hand a note to the recent editions of Pereira's Materia Medica (ii., p. 175) states that there is reason to believe that a gum resin like asafœtida is obtained from other species of Ferula. *Ferula Persica* has been described by Dr. Pope (*Phil. Trans.*, lxxv.) as the true asafœtida plant, and the Edinburgh College has admitted it as being, probably, one source of asafœtida. Michaux sent its fruits from Persia as asafœtida. That it does really yield that resin seems furthermore probable from the strong smell of that drug which pervades the whole plant.

In the Botanical Magazine (for March 1860) the writer observes, "Referring to our " herbarium we find various plants (varieties, genera, or species) all yielding the " asafœtida of commerce, or an entirely similar gum resin; (1) Dr. Falconer's plant " (leaves, fruit, and root from Tibet); (2) a very similar one collected by Drs. Falconer " and Thomson in the southern damp valleys of the same mountain (and elsewhere in " Kashmire), in whose northern dry valleys Falconer obtains his *Narthex*, also by " Dr. Thomson in Piti (Tibet); (3) a flowering specimen gathered in Turkestan by " Dr. Lord, 19 April 1838, and given to Dr. Falconer; it is much injured by insects; " (4) leaves and roots of a quite similar plant sent by Dr. Stocks from Doobund " in Beloochistan, as certainly the asafœtida of commerce; (5) another similar plant "from the banks of the Zenderad, in the Baktiyari mountains of Persia, collected by "the late W. Loftus (June 7, 1852), of which excellent specimens are in the British "Museum; (6) the *Scorodosma* of Bunge, of which we know the fruit, root, and "stems, but have not seen leaves; lastly we have imperfect fragments of oriental "Umbellifers from Aucher-Eloi and others, which may belong to some of the above." (*Hook. in Bot. Mag.*) Hence then we may conclude that the majority and weight of evidence is in favour of more than one, probably several, Umbelliferous plants yielding gum resins identical or similar to asafœtida, but of which the botanical names and characters are as yet uncertain.

Dr. W. Dymock recently contributed some observations upon the asafœtida in use in Bombay to the Grant College Students' Journal (Feb. 1874), in which he writes:— "Any European who has observed the asafœtida in use in Bombay must have "remarked how greatly it differs in appearance from that met with in shops in "Europe, and described in most books on Materia Medica. The asafœtida met with "in Bombay is in large masses enclosed in a cowhide, of the colour of treacle, and "consisting of the gum resin and sections of the root in about equal proportions. The "degree of hardness depends upon age; when fresh the consistence is that of yellow "wax. The odour is distinctly different from that of the European drug, and is "much more powerful. Expecting to find some notice of an article so entirely "different from that known in European commerce, I searched a number of works on "Materia Medica, but have only found it mentioned by Guibourt; it is therefore "evident that Bombay asafœtida is practically unknown in the European markets. "From reference to modern writers we obtain reliable information that there are two "or more plants which produce asafœtida."

To return to the Bombay drug, there are two kinds of genuine, and several kinds of adulterated asafœtida generally met with in the shops; the highest priced is known as Heera Hing, the juice of the plant. It is a liquid gum resin of the consistence and colour of thick treacle mixed with a small proportion of impurities, such as sand, fragments of the plant, &c., from which it can be easily separated by straining; if exposed to the air it hardens and becomes brittle, and semi-transparent; the odour is very powerful and pungent. This kind is much valued by the Banyas as a condiment. Ordinary asafœtida consists of the gum resin and slices of the root as already described. If genuine, the gum resin is opaque and dull, and the odour very powerful. The adulterated kinds consist of asafœtida mixed with common gum in various proportions; this admixture gives a brightness and transparency to the mass, which makes its detection a matter of no great difficulty. Sections of the root which is found mixed with Bombay asafœtida show the following peculiarities. A very irregular and confused arrangement of the different zones of which the root is composed. This is most marked in the large pieces. Large and peculiar lactiferous vessels disposed in groups of two, three, or more, sometimes near the centre of the root, sometimes near the periphery. A vascular system, composed of large spiral vessels, which are collected into bundles round the medullary portion of the root. Numerous cells, loaded with starch granules, having a large hilum in all parts of the root.

In European specimens of the root, which may sometimes be obtained from a mass of second sort asafœtida, there will be seen a perfectly regular arrangement of the zones, contrasting strongly with the Bombay specimens; they are, like them, remarkable for very large lactiferous vessels, but distributed symmetrically, the largest occupying the outer radius of the section; moreover, hardly any starch is seen, and none of the bundles of spiral vessels so prominent in the Bombay root.

The substance known as asafœtida in Europe is called Hingra in Bombay; it is not used as a condiment, and is considered quite a different article; it is exported to Europe. There is a great difference in price. Hingra is about Rs. 7 a Surat maund, and asafœtida about Rs. 45.

It is almost certain that asafœtida is produced by *Ferula Lehmanni*, Boiss, in the Aral-Caspian region, but to what extent it is collected, and its destination, is unknown. Probably also *Ferula alliacea*, Boiss, in Khorassan, &c. yields a portion.

IMPORTS of ASAFŒTIDA (Hingra) into the BOMBAY PRESIDENCY.

—	1869–70.	1870–1.	1871–2.	1872–3.
From Persian Gulf - -	lbs. 307,518	lbs. 22,130	lbs. 494,256	lbs. 535,360

EXPORTS of ASAFŒTIDA (Hingra) from the BOMBAY PRESIDENCY.

—	1869–70.	1870–1.	1871–2.	1872–3.
To	lbs.	lbs.	lbs.	lbs.
Foreign parts - -	—	—	126,568	348,480
Other Presidencies -	—	—	504	48,958
	—	—	—	—

EXPORTS of ASAFŒTIDA (Hing and Hingra) from the BOMBAY PRESIDENCY.

—	1869–70.	1870–1.
To	lbs.	lbs.
Foreign parts - -	172,088	24,702
Other Presidencies -	111,128	106,332
	—	—

IMPORTS of ASAFŒTIDA (Hing) into the BOMBAY PRESIDENCY.

—	1869–70.	1870–1.	1871–2.	1872–3.
From	lbs.	lbs.	lbs.	lbs.
Aden - - -	—	—	—	—
Persian Gulf - -	234,050	229,775	132,720	345,072
Soumeanee - -	98	—	—	—
Total - -	234,148	229,775	132,720	345,072
From				
Bengal Presidency -	—	—	—	448
Madras do. -	3,304	1,008	9,072	896
Scinde - - -	39,655	27,548	27,216	30,688
	—	—	—	—

EXPORTS of ASAFŒTIDA (Hing) from the BOMBAY PRESIDENCY.

—	1869-70.	1870–1.	1871–2.	1872–3.
To	lbs.	lbs.	lbs.	lbs.
Foreign ports - -	—	—	2,353	1,218
Other Presidencies -	—	—	148,582	147,349
Total - -	—	—	—	—

Bibliography of Asafœtida.

Christison, Dispensatory, p. 189. London, 1848.
Pereira, Materia Medica, Vol. ii. pt. 2, p. 174. London, 1857.
Royle, Materia Medica, pp. 437. London, 1865.
Kœmpfer Amœnitates Exoticæ, 535.
Lindley, Flora Medica, pp. 45–46. (1838.)
Lindley, in Botanic Register, August 1839.
Stephenson and Churchill, Medical Botany, iv. 169. London, 1831.
Nees and Ebermaier, Handbuch, ii. 55.
Lieut. Burnes' Travels, ii. 243.
Ainslie, Materia Medica, i. 21. London, 1826.
Pharmacopœia of India, pp. 101.
Fee, Hist. Nat. Pharm. ii. 199.
Guibourt, Hist. Nat. des Drogues, iii. 220. Paris, 1850.
Smith's Chinese Materia Medica, under *Asafœtida*. 1871.

Botanical Magazine, pl. 5168, March 1st 1860.
Balfour, in Edinburgh new Philosophical Journal, July 1858.
Birdwood's Catalogue, pp. 41, 268. Bombay, 1862.
O'Shaughnessy's Bengal Dispensatory, p. 362. Calcutta, 1841.
Balfour, in Transactions of the Royal Society of Edinburgh, vol. xxii. t. 20, 21.
United States Dispensatory, p. 145. Philadelphia, 1867.
Stewart's Punjab Plants, p. 106. Lahore, 1869.
Dictionaire de Matiere Medicale, i. p. 398. Paris, 1774.
Geiger Handbuch der Pharmacie, ii. p. 1342. Heidelborg, 1840.
Woodville, Medical Botany, i. p. 111. London, 1832.
Nees v. Esenbeck, Plantæ Officinales, 18 t. 16. Dusseldorf, 1828.
Powell, Punjab Products, p. 404. Roorkee, 1868.
Davies, Report on the trade and resources of the Countries on the N.W. boundary of British India. Lahore, 1863.
Pharmaceutical Journal, 1st ser. i. 461, i. 605, xii. 44, x. 351. 2nd ser. viii. 607, v. 583, ix. 588.
Buchner's Rep. No. xvi. (1850) p. 83.
Paulus Ægineta, and Commentary, iii. 337. London, 1847.
Pomet, History of Drugs, p. 189. London, 1737.
Flückiger and Hanbury, Pharmacographia, p. 281.

Dorema ammoniacum, *Don.*

Don in Linn. Trans. xvi. p. 599.

Disemeston gummiferum, *Jaub. and Spach Illus. Plant Orient.* i. t. 40.

A glaucous green plant about 7 feet high. Root perennial stem about 4 in. in circumference at the base. Leaves large, petiolate, somewhat bipinnate 2 feet long; pinnæ in three pairs; petioles downy, sheathing at the base; umbels proliferous racemose, partial ones globose, on short stalks, often arranged in a spiked manner; involucre general or none. Petals white, stamens and styles white, ovaries buried in wool, fruit naked.

Hab.—Persia.

Synonyms according to Moodeen Shereef:—

Ush-shaq, Ush-shaj. *Ar.*	Gama-nàyakam. *Tam.*
Ush-shah, Kilyanis. *Per.*	Gama-nàyakam. Tel.

Dr. Royle disposes of Ammoniacum in his Illustrations by the very brief remark that it is called *Ooshk* and *Ooshak*, is produced by *Dorema ammoniacum* of Don, in the province of Irak, where the plains are dry, gravelly, and exposed to an ardent sun (p. 231.)

The gum resinous exudation from the stem occurs in tears or masses; the tears, from two to eight lines in diameter, pale cinnamon-brown, breaking with a smooth, shining, opaque, white surface; the masses, composed of agglutinated tears, hard and brittle when cold, but readily softening with heat; has a faint odour, and a bitter, acrid nauseous taste; rubbed with water it forms a milky emulsion. Its activity resides in a volatile oil and a resin. (*Ind. Pharm.* 103.)

Much uncertainty long existed as to the ammoniac plant. It was generally believed to be a *Ferula* till Willdenow raised from some seeds mixed with the gum resin found in the shops, a plant which he ascertained to be a *Heracleum* and named *Heracleum gummiferum*, under the impression that it must be the source of the medicine. On this authority the plant was adopted by the British Colleges; Willdenow expressly acknowledged that he could not procure from it any gum resin, but ascribed the result to the influence of climate. The *Heracleum* however did not correspond exactly with the representations given of the ammoniac plant by travellers, and Sprengel ascertained that it was a native of the Pyrenees, and never produced gum. Mr. Jackson in his account of Morocco, imperfectly described a plant of that country, supposed to be a *Ferula*, from which gum ammoniac is procured by the natives. This plant was ascertained by Dr. Falconer to be *Ferula Tingitana*, and its product is thought to be the Ammoniacum of the ancients, which was obtained from Africa, but this is not the drug now used under that name, which comes exclusively from Persia. M. Fontanier, who resided many years in Persia, saw the ammoniac plant growing in the province of Fars, and sent a drawing with specimens of it to Paris; from these it was inferred to be a species of *Ferula*, and Merat and De Lens proposed for it the name, originally given to it by Lemery, of *Ferula ammonifera*. It was subsequently however ascertained from specimens obtained in Persia by Colonel Wright and examined by Dr. David Don that it belonged to a genus allied to Ferula, but essentially different, which was named by Dr. Don *Dorema*. It was described in the Linnean Transactions (vol. xvi.) under the name of *Dorema ammoniacum;* this is now acknowledged by the official authorities.

The ammoniac plant grows spontaneously in Farsistan, Irauk, Chorassan, and other Persian provinces. Dr. Grant found it abundantly in Syghan near Bameean on the north west slope of the Hindoo Coosh mountains. It attains the height of 6 or 7 feet, and in the spring and early part of summer abounds in a milky juice, which flows out upon the slightest puncture. From the accounts of travellers it appears that in the month of May the plant is pierced in innumerable places by an insect of the beetle kind. The juice exuding through the punctures concretes upon the stem, and when quite dry is collected by the natives. M. Fontanier states that the juice exudes spontaneously and that the harvest is about the middle of June. According to Dr. Grant the drug is collected in Syghan like asafœtida from the root of the plant; the gum resin is sent to Bushire, whence it is transmitted to India, chiefly to Bombay; a small portion is said to be taken to the ports of the Levant, and thence distributed. The name of the drug is thought to have been derived from the temple of Jupiter Ammon in the Libyan Desert, where the ammoniac of the ancients is said to have been collected, but Dr. Don considers it a corruption of Armeniacum, originating in the circumstance that the gum resin was formerly imported into Europe through Armenia. (*U. S. Dispens.* p. 105.)

Pereira was unable to ascertain when the Persian Ammoniacum first came into use. As the Greeks and Romans make no mention of it, they were probably unacquainted with it. Avicenna does not mention the origin of his Ammoniacum (*assaoh,* Ar.) The ammoniac (*eschak,* Ar.) of Abu Mansur Mowafik, an ancient Persian physician, who wrote about 1055 A.D. was doubtless of the Persian kind, as was also the ammoniac (*derukht ushak*) of Beva Ben Khuas Khan, A.D. 1512. The Arabic terms (*assach, ushak,* and *oshac*) by which the three last-mentioned authors designate ammoniac, resemble the name by which the Ammoniacum plant is now known in Persia, hence we infer they all referred to the same object. (*Pereira, Mat. Med.* ii. 185.)

Buhse found the plant which yields gum Ammoniacum at the foot of the mountains of Kuni Nirehm. He considers that it is not the *Dorema ammoniacum,* but the *Dorema* which grows on sandy and rocky elevations from 3,000 to 8,500 feet above the level of the sea. (*Pharm. Journ.* xi. p. 526.)

This is alluded to in Royle's Materia Medica, and remarked that Buhse's plant was called by the natives *Weshack,* and he concluded it to be *Dorema Aucheri,* but he saw only the leaves, which were probably those of *D. ammoniacum.* The gum may, however, be yielded also by other allied species. (*Mat. Med.* 445.) In a note to the same work is added that "the energetic traveller Aucher-Eloy visited among other places Yezdikhast and obtained fragments of a plant which has been named *Diserneston gummiferum* by Jaubert and Spach in Illustr. Pl. Orient., who state that it is allied to *Siler* and to *Agasyllis* and is hence placed in the tribe *Sileridæ* (*Walpers' Repert.* ii. 939.) It appeared to Dr. Royle to be only *Dorema Ammoniacum* described from imperfect specimens.

The smell of ammoniac is peculiar, and stronger in the mass than in the tears. The taste is slightly sweetish, bitter, and somewhat acrid. The spec. grav. is 1·207. When heated the gum resin softens and becomes adhesive, but does not melt. It burns with a white flame, swelling up and emitting a smoke of a strong resinous, slightly alliaceous odour. It is partly soluble in water, alcohol, ether, and alkaline solutions. Triturated with water it forms an opaque milky emulsion which becomes clear upon standing. The alcoholic solution is transparent, but is rendered milky by the addition of water. Bucholz obtained from 100 parts of ammoniac 22·4 parts of gum, 72·0 of resin, 1·6 of bassorin, and 4·0 of water including volatile oil and loss. Hagen succeeded in procuring the volatile oil in a separate state by repeated distillation with water. It has a penetrating, disagreeable odour, and a taste at first mild, but afterwards bitter and nauseous. The resin of ammoniac is dissolved by alcohol, and by the fixed and volatile oils; but it is divided by ether into two resins, of which one is soluble, the other insoluble in that menstruum. (*U. S. Dispens.* 106.)

The museum specimens are from Calcutta. In the reference collection is a sample in tears from Seistan, and one or two peculiar varieties and forms of mixed Ammoniacum in lump from the London market.

Dorema aureum, *Stocks.*

Stocks in Hook. Journ. Bot. (1852), p. 149.

Stem striate, nearly leafless, radical leaves at first clad with a short scattered pubescence, at length smooth, ternato-tripinnatisect, ultimate segments lanceolate, decursive, becoming parallel, more or less lobed; panicles very much branched, pubescent; mericarp elliptical, smooth; juga filiform, entire.

Hab.—Doobund, and elsewhere in Upper Beloochistan.

Stocks states that it yields a bitter white gum much like the Ammoniacum of the shops. The plant often stands six feet high, and is very conspicuous from the golden hue of the loose and much branched panicles.

Ferula orientalis, *L.*

Tournef. Voy. ii. 379. *Boiss. Fl. Orient.* p. 986.

Stem terete, branched, leaves five times pinnate; leaflets multifid setaceous, apparently puberulous when examined by a lens; sheaths large cucullate, involucra wanting. Fruit obovate oblong, mericarps convex, commissural face slightly concave, margin narrow.

Hab.—Armenia. E. Caucasus, &c.

In Royle's Materia Medica it is stated that the gum resin of this species, as obtained by Dr. Stocks in Scinde, is wholly unlike any product known in English commerce. It would be well to determine if this plant is really found in Scinde, or a similar and allied species.

IMPORTS of GUM AMMONIACUM into the PRESIDENCY of BOMBAY.

—	1869–70.	1870–1.	1871–2.	1872–3.
From Persian Gulf - -	cwt. 327	cwt. 520	cwt. 164	cwt. 1671

EXPORTS of GUM AMMONIACUM from the PRESIDENCY of BOMBAY.

—	1871–2.
To United Kingdom - -	cwt. 453

Bibliography of Ammoniacum.

Pereira, Materia Medica, vol. ii. pt. 2, p. 184.
Ainslie, Materia Medica, i. 159. London, 1826.
Pharmaceutical Journal, xi. p. 526.
Don in Linnean Transactions, xvi. p. 605.
Jackson, Account of the Empire of Morocco, 3rd ed. p. 156.
Braconnot, Ann. de Chimie, lxviii. p. 69.
Gmelin, Handb. d. Chem., ii. 624.
Jaubert & Spach, Illustr. Plant. Orient., i. t. 40.
Pharmacopœia of India, p. 103.
United States Dispensatory, p. 105. Philadelphia, 1867.
Stocks in Hooker's Journ. Bot., ii. p. 149. London, 1852.
Guibourt, Hist. Nat. des Drogues, iii. 224. Paris, 1850.
Royle, Materia Medica, p. 444. London, 1865.
Lindley, Flora Medica, p. 47. (1838.)
O'Shaughnessy, Bengal Dispensatory, p. 364. Calcutta, 1841.
Dictionaire de Matiere Medicale, i. p. 277. Paris, 1774.
Geiger, Handbuch der Pharmacie, ii. 1,337. Heidelberg, 1840.
Powell, Punjab Products, p. 404. Roorkee, 1868.
Pharmaceutical Journal, 1st ser. i. 578; ii. 773; xi. 526.
Paulus Ægineta, and Commentary, iii. 38. London, 1847.
Christison, Dispensatory, p. 97. London, 1848.
Pomet, History of Drugs, p. 192. London, 1737.
Flückiger and Hanbury, Pharmacographia, p. 288.

Ferula galbaniflua, *Buhse.*

Buhse in Boiss. Fl. Orient. p. 988.

Opoidea galbanifera. *Don in Linn. Trans.* (?)

Stem thick, high, straight, naked, verticillately branched above; leaves rather greyish downy, quadripinnatisect, primary and secondary divisions long-petiolate, segments ovate, cut in very short crowded divaricute linear, entire or trifid limbs, cauline leaves reduced to acute deciduous oblong vaginæ. Umbels 6–12 radiate, involucel obsolete, pedicels very short incrassated, petals yellow, smooth, fruit oblong or elliptic, vittœ solitary, commissure evittate.

Hab.—Persia.

The native synonyms are—

Oinnah. *Ar.*
Barazd, Bárazhd, Bérazd. *Pers.*
Baríjá, Barija-gónd. *H. Duc.*

Of this gum resin Royle observes (*Illustr.* p. 231), "Galbanum in Persian works has " *barzud* given as the Arabic, *bireeja* as the Hindustanee, with *Rhulyan* and *metonioon* " as the Greek names (evident corruptions of χαλβανη and μετωπυον owing to errors in " the reading of the diacritical points). *Kinneh* and *nafeel* are stated to be the names " of the plant, which is described as being jointed, thorny, and fragrant. Neither its " locality nor nature are well known, but it is probable, as suggested by Mr. Don, " that the plant named by him *Galbanum officinale*, from some seed found on the " drug may be a native of some remote and inaccessible parts of Syria; perhaps of " some of the northern parts of Persia or of Arabia."

The gum resin occurs in irregular tears about the size of a pea, usually agglutinated with masses of a greenish yellow colour, translucent, having a strong disagreeable odour, and an acrid bitter taste. The two latter properties are due to the presence of a volatile oil.

At one time it was supposed that this gum resin was the product of *Bubon galbanum*, an umbelliferous plant of the eastern coast of Africa. It has also been referred to the *Ferula ferulago* of Linnæus, the *Ferula galbanifera* of Lobel, which inhabits the coasts of the Mediterranean, and is also found in Transylvania and the Caucasus. But no part of either of these plants has the odour of galbanum, and it is therefore scarcely probable that they yield the drug. Mr. Don having found the seeds taken from a parcel of galbanum to belong to an undescribed genus of umbelliferous plants, and concluding that they came from the same source as the gum resin itself gave the title of *Galbanum* to the new genus, and named the species *Galbanum officinale*. This was rather hastily adopted by the London College, as it is by no means certain that the same plant produced the seeds and the gum resin. Specimens of a plant were received in England from Persia having a concrete juice adhering to them, which was taken by Dr. Lindley for galbanum, and that botanist finding that the plant belonged to an undescribed genus named it *Opoidea*, with the specific name *galbanifera*. Dr. Pereira, however, found the substance not to be galbanum, and this supposed origin of the drug, therefore, though admitted as probable by the Edinburgh College, and recognised by the Dublin, must be considered as more than doubtful. A German traveller, F. A. Buhse, who has resided in Persia, states that in 1848 he met with the galbanum plant on the declivities of the Demawend, near the southern coast of the Caspian. He saw the gum resin exuding spontaneously from the plant, and was informed by the natives that the drug was collected from it. The plant is a *Ferula*, and closely resembles the *Ferula erubescens* of Boissier, if not identical with it. (*Pharm. Cent. Blatt.*, March 17, 1852, p. 206.) Galbanum is said to be obtained by making incisions into the stem or cutting it off a short distance above the root. A cream coloured juice exudes, which concretes upon exposure to the air. A portion of juice also exudes spontaneously from the joints and hardens in the shape of tears. (*U. S. Dispens.* p. 401.)

This resin is very rare and scarcely known in the Punjab. Mr. Powell writes, " whenever I asked for 'barzad,' and 'barija,' I invariably got a pot full of the " semi-fluid sticky whitish 'ganda-barosa,' or crude turpentine of the *Pinus longi-* " *folia*." He describes three different samples of pine resins sent to him for galbanum.

In June 1848, Buhse, a German traveller in Persia, states that he met with the true galbanum plant. "He found it on the declivities of the Demawend. It is a " *Ferula*, from the stalks of which a liquid issues abundantly, by the odour and nature

" of which he immediately recognised galbanum, and his guides assured him moreover " that galbanum is gathered from this plant. The author had not distinctly determined his plant (which has since been done, and named *Ferula galbaniflua*)." It appears to differ from *Ferula erubescens* only by the absence of commissural vittœ, but as neither Aucher-Eloy nor Kotschy, who have both collected the *Ferula embescens*, make any mention of its yielding galbanum, the author is in doubt whether his plant be the same or a variety of it. Don's genus *Galbanum* and Lindley's *Opoidea* do not agree with the plant seen by Buhse, unless that both of these authors have made their descriptions from imperfect fruits, or that there exist other plants which yield galbanum. The plant which Buhse described is called in some parts of Persia *Khassuch*, and appears to be confined to certain districts of Persia. In the whole large district of the Elburs chain, from the south-east angle to the south-west angle of the Caspian Sea, it is only found in the neighbourhood of the Demawend, but here at an elevation of from 4,000 to 8,000 feet, and even on the declivity of the top of the Demawend. It exists neither on the mountains of Talysch nor in the districts of Karadagh and Tabris. It is said to re-appear on the Mount Alwend, near Hamadan, and in the neighbourhood of the great salt desert. Near Hamadan Aucher-Eloy has gathered his *Ferula erubescens*, and this supports the supposition that the author's plant is the same. In the salt desert itself Buhse did not meet with it again. The inhabitants of the Demawend collect the gum-resin, which issues spontaneously from the lower part of the stalk, they do not make incisions in the plant; but it is not at this place that the galbanum is collected for commercial purposes. When fresh the gum resin is white like milk, liquid, and somewhat glutinous. In the air it soon becomes yellow, elastic, and finally solid. The odour is rather strong, unpleasant, and similar to that of our commercial galbanum. (*Pharm. Journ.* 1st ser. xi. p. 577.)

The odour of galbanum is peculiar and disagreeable, its taste bitterish, warm and acrid. Its sp. gr. 1·212. Triturated with water it forms an imperfect milky solution, which on standing deposits the greater portion of what was taken up. Wine and vinegar act upon it in a similar manner. Alcohol dissolves a considerable proportion, forming a yellow tincture, which has the smell and taste of galbanum, and becomes milky with water, but affords no precipitate. In dilute alcohol it is wholly soluble with the exception of impurities. Ether dissolves the greater portion. Pelletier found in 100 parts 66·86 parts of resin, 19·28 of gum, 6·34 of volatile oil, including the loss 7·52 of wood and impurities, with traces of supermalate of lime. A small proportion of bassorin was found by Meissner. By distillation at the temperature of about 250° F. the volatile oil is obtained of a fine indigo blue colour which it imparts to alcohol. Procured by distillation with water it is colourless, and becomes yellowish by age. It is lighter than water. (*U.S. Dispens.* p. 402.)

Russian or Persian Galbanum

According to Ludewig a gum resin, designated as Persian galbanum, is received in Russia by the way of Astracan or Orenburg, and is the kind used in that country. It comes enclosed in skins, and is in masses of a reddish-brown colour with whitish streaks, of a disagreeable odour, somewhat like that of asafœtida, and of an unpleasant bitter resinous taste. It is so soft as to melt with a slight elevation of temperature. It differs from common galbanum in its odour, in its colour which is never greenish, and in the absence of tears, and is probably derived from a different plant. It abounds in impurities. (*U.S. Dispens.* p. 402.)

Bibliography of Galbanum.

Pereira, Materia Medica, vol. ii. p. 2, p. 188. London, 1857.

Don, in Linnean Transactions, xvi. p. 603.

Lindley, in Botanical Register, Aug. 1, 1839, p. 65.

Solly, proceedings of Committee of Commerce and Agriculture of Royal Asiatic Society, p. 144. London, 1841.

Pelletier, Bullet de Pharm. iv. 97.

Pharm. Centr. Blatt, March 17, 1852, p. 206.

United States Dispensatory, p. 401. Philadelphia, 1867.

Royle, Materia Medica, p. 446. London, 1865.

Lindley, Flora Medica, pp. 47, 51. (1838.)

Ainslie, Materia Medica, i. 143, 601. London, 1826.

Pharmacopœia of India, p. 104.

Guibourt, Hist. Nat. des Drogues, iii. 226. Paris, 1850.

O'Shaughnessy, Bengal Dispensatory, p. 367. Calcutta, 1841.
Graves, Hort. Med., under "Galbanum."
Thomson's Elements of Materia Medica, ii. p. 165.
Birdwood Catalogue, pp. 42, 268.
Dictionaire de Matiere Medicale, iv. p. 13. Paris, 1774.
Geiger, Handbuch der Pharmacie, ii. 1834. Heidelberg, 1840.
Woodville, Medical Botany, i. p. 98. London.
Nees v. Esenbeck, Plantæ Officinales, vii. t. 10. Dusseldorf, 1828.
Powell, Punjab Products, p. 404. Roorkee, 1868.
Pharmaceutical Journal, 1st ser. xi. 577; 2nd ser. iii. 428.
Paulus, Ægineta and Commentary, iii. 398. London, 1847.
Christison's Dispensatory, p. 454. London, 1848.
Pomet, History of Drugs, p. 189. London, 1737.
Flückiger and Hanbury, Pharmacographia, p. 285. London, 1874.

Opoponax chironium, *Koch.*

Plant 6–7 feet high, resembling the parsnip, perennial; root thick; stem rough; leaves bipinnatisect; segments unequally cordate, crenate, obtuse; umbels compounded of many rays, general and partial, involucre few-leaved. Flowers yellow. Margin of the calyx obsolete. Petals roundish, entire, rolled inwards with a rather acute lobe. Stylopodium broad thick. Styles very short. Fruit flattened at the back with a dilated convex margin, mericarps with three dorsal filiform, very thin ridges, and no distinct lateral ones. Vittœ 3 to each channel, 6–10 to each commissure; seed smooth.

Hab.—S. France. Italy. Sicily. Croatia, and Greece.

The following are given as synonyms in the North-west Provinces of India—

Jaushir, Gaushir, Jawashir, Hashi, Jahoshi, Hoshi. *N.W.P.*

According to Dioscorides, this gum resin is obtained by incisions into the root; a milky juice exudes, which by drying becomes yellow, and forms Opoponax.

It occurs in irregular yellowish red lumps or in reddish tears. It has an acrid bitter taste and an unpleasant odour. Rubbed with water it forms an emulsion. (*Pereira Mat. Med.* ii. 192.)

Opoponax is described by Dioscorides as the produce of a plant of Bœotia and Arcadia, which has been identified with the present *Opoponax chironium*, referred by Sprengel to the genus *Ferula.* It is found also in the open fields of the south of France, of Italy, Sicily, and Greece, and according to Merat and De Lens also in Syria and the East. Dodœns first grew this plant from seeds found attached to pieces of Opoponax, and he states that when wounded in warm weather, especially near the root, a juice exudes which concretes into a gum resembling Opoponax. But there is no proof that this plant yields the Opoponax of commerce, which reaches India either from the Persian Gulf or the coast of Arabia, and is called *juwa sheer*, or the milk of *juwa*, as it was by Serapion. Opoponax is imported into England from Turkey; in the time of Matthioli it was obtained from Alexandria. (*Royle, Mat. Med.* p. 436.)

Dr. Porter Smith says that it is probable this drug has been known to the Chinese at some time or other, judging from the plate in the Pen Tsau, its association with other Umbelliferous plants, and other considerations.

Ainslie remarks that he never had found this gum resin in any of the bazaars of lower India, it has however a place in the Ulfaz Udwiyeh and is of course known in the more northern tracts of Hindoostan (*Mat. Med.* vol. i. p. 280.) O'Shaughnessy on the other hand says that it is found in all the bazaars of Bengal, and is even exported from this country to Europe.

The Makhzan-ul-adwiya observes that this gum comes from a village called Máh near Ispahan, where also "sakbinaj" is produced. The author describes the plant as "having abundance of close set stalks, somewhat tall and covered with a white down." Its leaves are described as like those of the fig tree. The flower head "kubah" is like that of the "Shibt" (*Anethum sowa*), the flowers being yellow and scented. The seeds are black like anise, and have an aromatic smell. The author goes on to state that the gum is obtained by taking up the root at the time when the plant begins to sprout, and breaking it open and allowing the juice that exudes to concrete on leaves placed below. An inferior kind, he adds, is made by mixing up wax and ammoniacum by way of adulteration, with a little of this gum. Powell believes that the origin of the Punjab druggists "jaushir" is Persian, and the countries to the south as far as Beloochistan. The Persian origin of the names and the coincidence of the Beloochistan name and the Shirazi names "hushi" and "jahushi" are confirmatory of the supposition. In Lahore the gum sells at 6 annas a seer. At Amritsur he recently obtained a sample, which the owner assured him had come direct from Bushire, and was of superior quality. It was in quite small tear-like pieces of a dark tawny reddish yellow outside, and whitish inside, being much smaller and darker than the ordinary samples. (*Powell, Punj. Prod.* p. 403.)

Opoponax has been analysed by Pelletier and found to contain—

Resin - - - - - - -	42
Gum - - - - - - -	83·4
Starch - - - - - - -	4·2
Extractive - - - - - - -	1·6
Wax - - - - - - -	0·3
Malic acid - - - - - - -	2·8
Lignin - - - - - - -	9·8
Volatile oil, traces of caoutchouc and loss -	5·9

It is fusible at 122° F. Soluble in alkalies, alcohol, and ether. The alkaline solution is reddish, the resin is precipitated from it by hydrochloric acid in the form of yellow flocks. Nitric acid acts freely on the resin. (*Pereira Mat. Med.*)

Ferula hooshee, *Royle.*

Royle, Illustrations Himalayan Botany, p. 231.
Hab.—Beloochistan.
Native synonyms :—

Hooshee.

Dr. Royle states in his Illustrations (p. 231) that "Dr. Lindley has some seeds, "called 'hooshee' sent by Mrs. Macneil from the hills of Beloochistan, which he "informs me appears to be Opoponax." O'Shaughnessy writes of it that it resembles Asafœtida in size and appearance, and yields a gum like the Opoponax of European shops.

Bibliography of Opoponax.

Pereira, Materia Medica, vol. ii. pt. 2, p. 191. London, 1857.
Royle's Illustrations, p. 231.
Royle, Materia Medica, p. 436. London, 1865.
Lindley, Flora Medica, pp. 44, 46. (1838.)
Stephenson and Churchill, Medical Botany, iii. pl. 98. London, 1831.
Ainslie, Materia Medica, i. 281. London, 1826.
Guibourt, Hist. Nat. des Drogues, iii. 229. Paris, 1850.
O'Shaughnessy, Bengal Dispensatory, p. 361. Calcutta, 1841.
United States Dispensatory, pp. 1570. Philadelphia, 1867.
Graves, Hortus Medicus, pp. 195.
Cox's Medical Botany, pl. 40. London, 1821.
Birdwood's Catalogue, p. 268.
Dictionaire de Matiere Medicale, v. p. 305. Paris, 1774.
Geiger, Handbuch der Pharmacie, ii. p. 1340. Heidelberg, 1840.
Woodville, Medical Botany, i. p. 122. London, 1832.
Nees v. Esenbeck, Plantæ Officinales, 17 t. 11. Dusseldorf, 1828.
Powell, Punjab Products, p. 402. Roorkee, 1868.
Christison's Dispensatory, p. 687. London, 1848.
Pomet, History of Drugs, p. 191. London, 1737.
Flückiger and Hanbury, Pharmacographia, p. 291.

Sagapenum, *Ferula sp.*

The following are amongst the synonyms :—

Sugbeenuj, Sakbinaj. *Ar.*	Sagafioon. *Pers.*
Kundel. *H.*	Eesus. *Mahr.*

Sagapenum, *sugbeenuj* of the Arabs, and to which *sugafiyoon* is assigned as the Greek name in Persian authors, is probably a product of Arabia, as it is imported thence into India, and into England from Alexandria; it has been supposed to be produced by *Ferula Persica*, but the subject is worthy of investigation.

Thus wrote the late Dr. Royle, and still but little more knowledge or certainty exists, and still the subject is "well worthy of investigation by travellers."

Two kinds of Sagapenum are met with, which are thus characterised in Pereira's Materia Medica.

Sagapenum in tears consists of masses made up of agglutinated brownish-yellow semi-transparent tears, and resembling galbanum, but having a darker colour and a more alliaceous odour.

Soft Sagapenum occurs in soft tough masses, in which no distinct tears are distinguishable. When heated on the point of a knife in the candle, sagapenum gives out a much more aromatic and agreeable odour than galbanum. It has a hot and acrid taste.

Sagapenum like opopónax, has been known to us since the time of the Greeks. Dioscorides describes it as the produce of a Ferula growing in India, and we have no more recent information. It is the *sukbeenuj* of the Arabs who gave *sagafioon* as its Greek name. It reaches India from the Persian Gulf or the coasts of Arabia. It is difficult by price currents and commercial reports to distinguish the routes which small articles of commerce pursue. It is imported into Europe from the Levant and also from Alexandria. It is probably a product of Persia. Willdenow was of opinion that it was produced by *Ferula Persica,* which Oliver thought produced Ammoniacum, and Dr. Hope that it yielded asafœtida; and which probably does yield some kind of fœtid gum resin, as Michaux sent its seeds from Persia as those of asafœtida, but it is preferable to leave this as a subject for investigation, than to stop inquiry by promulgating imperfect information. (*Royle, Mat. Med.* p. 437.)

The Makhzan-ul-adwiya says that this gum resin is found near Ispahan at a village called Máh. In the Punjab it is extremely difficult to obtain and very costly. With great difficulty Mr. Powell obtained a small piece for two rupees from a Pandit who had travelled much in Khorasan and Bukhara. This piece was very like ammoniacum with a white waxy appearance and yellow outside, part of which had been removed. The druggists if asked for "sakbinaj," as none of them have the real drug, produce opoponax or ammoniacum. The sample exhibited at the Lahore Exhibition from Amritsar was evidently not "sakbinaj" but a dark massive gum resin, full of little sticks and impurities; which when broken exhibited, when it powdered in breaking, a green colour, it had a bitter taste, and a somewhat alliaceous smell when broken. (*Powell Punj. Prod.* p. 404.)

Pelletier's analysis of sagapenum gives,—

Resin	54·26
Gum	31·94
Volatile oil and loss	11·80
Bassorin	1·00
Malate of lime	0·40
Peculiar matter	0·60

Whereas Brandes's analysis is as follows:—

Resins	50·29
Gum with calcareous salts	32·72
Volatile oil	3·73
Bassorin	4·48
Malate and phosphate of lime	1·12
Impurities	4·30
Water	4·60

Bibliography of Sagapenum.

Pereira, Materia Medica, vol. ii. pt. 2, p. 183. London, 1857.
Gmelin, Handbuch and Chemie, ii. 625.
Royle, Materia Medica, p. 437. London, 1865.
Lindley, Flora Medica, p. 46. (1838.)
Ainslie, Materia Medica, i. 358. London, 1826.
Guibourt, Hist. Nat. des Drouges, iii. 224. Paris, 1850.
O'Shaughnessy, Bengal Dispensatory, p. 363. Calcutta, 1841.
United States Dispensatory, p. 1594 Philadelphia, 1867.
Dictionaire de Matiere Medicale, vii. p. 10. Paris, 1774.
Geiger Handbuch der Pharmacie, ii. 1345. Heidelberg, 1840.
Powell, Punjab Products, p. 403. Roorkee, 1868.
Paulus Ægineta, and Commentary, iii. 322. London, 1847.
Christison's Dispensatory, p. 813. London, 1848.
Pomet, History of Drugs, p. 190. London, 1737.

Sarcocolla.

The native synonyms are—

Goojur. *Mahr.*
Unzeroot. *Ar.*
Kunjudeh. *Pers.*

Occurs in small rounded tears or grains, from the size of coarse sand to that of small peas, yellowish or straw colour, mixed with tears of a darker reddish colour, rather friable, dull.

This subviscid, sweetish, and somewhat nauseous gum resin, says Ainslie, is but rarely met with in India, and what is found is brought from Persia or Arabia; it is seen in grains of a pale yellow colour. It is an article of the Mogul Materia Medica, and is well known to the Arabians, who suppose it to have virtues applied externally, in agglutinating wounds (hence its Greek name σαςξ κολλα), and accordingly place it amongst their Zabisaut kerouh. Some of the more ancient Arab writers, such as Mesue, seem to have considered it as, in a certain degree, cathartic, though it does not appear to have been in much repute. The shrub which yields the sarcocolla by spontaneous exudation, is a native of Ethiopia; having leaves acuminate and smooth, with red terminating flowers.

"Sarcocol," says Pomet, "is a gum that flows from a little prickly shrub whereof the leaves are like Palta sena, of a white colour tending to a yellow. Almost all authors say that these shrubs grow in Persia, but two friends of mine at Marseilles wrote to me in July 1692 after this manner, that sarcocola was a gum that is gathered in the deserts of Arabia, the tree being little and very thorny; choose that in tears and grains of a white colour, inclining to yellow or red, of a sweetish taste attended with an unpleasant bitterness. This gum is wonderful in its kind, in that it flows from a tree without incision, and by incision in tears or drops of different colour and bigness, being white, yellow, and red, and when they are dry are granulated or in grains as they are brought to us from Marseilles. There is besides another sort of sarcocol, which is in the brown lump and looks like a composition, but I take it to be nothing but the damaged gum that ought to be thrown away, and that is very full of dirt, and other filth, to which it is much subject."

Lemery has very similar observations. He says Sarcocolla or Flesh-glue is a granulated gum, in very small bits that are spongy, of a yellowish colour tending to white, resembling the broken pieces or remains of a gum, or of Frankincense that have been grossly powdered. We have it brought us from Arabia, where, as some say, it comes from a thorny shrub whose leaves are something like senna. Sarcocol is of such a brittle nature that it moulders down into a gross powder only by shaking or handling it, wherefore it is commonly brought over to us in small grains, not much bigger than poppy seeds, yet sometimes we meet with some that are bigger and thicker. They are of several colours, viz., white, brown, and red, of a bitter taste, with a certain nauseous sweetness, and being chewed between the teeth they become tough; being held to the flame of a candle they first boil or fry, and then send forth a clear flame.

This substance is still referred by most authorities to a species of *Penæa*, usually to *Penæa mucronata*, whilst Endlicher is of opinion that it is not the produce of this order at all, and the late Dr. Lindley was disposed to regard it as of Umbelliferous origin. When met with in India at all it seems to occur in the channels of the Persian trade from the north, which favours the supposition that it is derived from Persia, whereas the assumption that it is yielded by a plant, native of the Cape of Good Hope, seems wholly gratuitous. For the present, at any rate, it must be confessed that the evidence is so weak that we are justified in the conclusion that both source and country are unknown.

According to Pelletier this gum resin contains—

Pure sarcocolline	65·30
Gum	4·60
Gelatinous matter	3·30
Woody fibre	26·80

Sarcocolline is half transparent, crystalline, brittle like gum, soluble in 40 parts of cold water, and 25 of boiling water; soluble also in alcohol. Odour weak and peculiar, taste saccharine and slightly bitter. (*O'Shaughnessy Disp.* p. 427.)

Bibliography of Sarcocolla.

United States Dispensatory, p. 1,596. Philadelphia, 1867.
Paulus Ægineta, and Commentary, iii. 327. London, 1847.
Guibourt, Hist. Nat. des Drogues, ii. 541. Paris, 1850.
O'Shaughnessy, Bengal Dispensatory, p. 427. Calcutta, 1841.
Ainslie, Materia Medica, i. pp. 380, 629. London, 1826.
Duncan, Edinburgh Dispensatory, p. 302.
Avicenna, Canon Med., lib. ii. tract ii. p. 36.
Pomet, History of Drugs, pp. 198. London, 1737.

Gardenia lucida, *Roxb.*

Roxb. Fl. Ind. Wight Icones, ii. t. 575.

Arborescent, unarmed, with resinous buds; leaves very shortly petioled, oblong, oval, or obovate, obtuse or with a short blunt point, glabrous, hard, shining, with simple parallel nerves and connecting veins; flowers somewhat terminal, solitary, rather shortly pedicelled (pedicels from ½ to 1 in. long); limb of the calyx with five long subulate divisions, sprinkled internally with erect short stout bristles; corolla hypocrateriform; tube long, glabrous striated; limb 5 partite, divisions obovate oblong, as long as, or a little shorter than, the tube, glabrous, stigma entire; berry drupaceous, even, oblong, crowned with the whole limb of the calyx, nut very hard, thick, and bony, with two parietal receptacles. (*Wight Icon.* t. 575.)

Gardenia gummifera, *Roxb.*

Roxb. Fl. Ind. Wight Icones, ii. 576.

Gardenia arborea. *Roxb.*

Arborescent, unarmed, with resinous buds; leaves sessile, from narrow elliptic oblong to ovate oblong, obtuse or very shortly and bluntly pointed, puberulous and slightly scabrous when young, afterwards shining, with simple parallel nerves; flowers terminal, 1–3 together, almost sessile; calyx densely puberulous and slightly scabrous; limb short, with 5 ovate acuminated divisions, corolla hypocrateriform; tube long, slender, widened at the mouth, sparingly pubescent; limb 5 partite, segments narrow oblong, more than half the length of the tube, almost glabrous; stigma clavate, entire, striated; berry drupaceous, even, oblong, crowned with the whole limb of the calyx, nut with 4 or 5 parietal receptacles. (*Wight Icon.* ii. 576.)

The native synonyms of the resin given by Moodeen Shereef are—

Dik mali. *H.*
Dika-mali. *Duk.*
Kumbai, Dika-malli. *Tam.*
Tella-manga, Chinaka-ringuva. *Tel.*
Dikke-malli. *Can.*
Kola-lakada. *Cing.*

Both these trees, which are common in many parts of India, yield a resinous exudation. (*Dika mali*, H.) It occurs in the form of irregular earthy looking masses, of a dull olive green colour, consisting of the resin, more or less mixed with bark, sticks, and other impurities. The twigs of the tree, coated and agglutinated with the exuding resin, are also sold in the bazaars. The odour is peculiar and offensive. It has been found useful in hospital practice in preventing the access of flies to ulcerated surfaces. According to Sprengel the resin of *Gardenia gummifera* constituted the *Cancamum* of the ancients, but this is doubtful. (*Ind Pharm. p.* 118.) Alluding to this, Dr. Birdwood in his Catalogue writes, "Can it be the χαγχαμον of Dioscorides, which (Lib. i., " cap. 23) he mentions as an Arabian product? It is imported into Bombay from " Arabia to the present day."

Ainslie, under the name of Cumbi gum, or *Cumbi pisin*, alludes to this as a strong smelling gum resin, not unlike myrrh in appearance, and possessing similar virtues; it is, however, far more active, and ought on that account to be prescribed in very small doses. As an external application it is employed, dissolved in spirits, in cleaning foul ulcers and preventing mortification. I have not been able to ascertain from what plant it is procured, or whence it comes. (*Mat. Med.* vol. ii. p. 89.)

The specimens in the Collection are derived from Mysore, Madras, and Central Provinces.

Bibliography of Dika-mali.

Pharmacopœia of India, pp. 118. London, 1868.
Walker, in Bombay Med. Phys. Trans., 1840, p. 70.
Long, in Journ. Agri. Hort. Soc., India, 1858, x. p. 10.

Sect. C.—FRAGRANT GUM RESINS.

These are distinguished from those in Sect. A. and B. rather by negative than positive qualities, inasmuch as they are not emulsive, and are not possessed of the fetid odour which characterises gum resins of the asafœtida kind. Most of the products included in this section have a decidedly pleasant odour, but this is more to be recognised in the last of the two subdivisions instituted.

Sub-div. α.—Bdellium kind.

This section includes gum-resins of the Myrrh and Bdellium kind, possessing in some but little odour, having a more or less aromatic or bitter taste, and usually of a darkish brown colour.

Balsamodendron Myrrha, *Ehr.*

Nees Düsseldorf, pl. 357.

Stem shrubby, arborescent; branches squarrose, spinescent. Leaves ternate; leaflets obovate, obtuse, obtusely denticulate at the apex, the lateral smooth. Fruit acuminate.

Bark pale ash grey, approaching white. Wood yellowish white, both it and the bark have a peculiar odour. Leaves on short stalks. Flowers unknown. Fruit ovate, smooth, brown, somewhat larger than a pea; surmounted at the base by a four-toothed calyx, and supported on a very short stalk. (*Pereira.*)

Hab.—Gison, on the borders of Arabia Felix.

The native synonyms given for Myrrh in Supplement to Indian Pharmacopœia are—

Mur, Murr. *Ar.*
Bol. *P. H. Duk.*
Vellaip-polam. *Tam.*
Balimtra-polam. *Tel.*
Bola. *Can.*
Bol. *B. & Guz.*
Gandha-rasaha, Rasa-gandhaha. *Sans.*
Balata-bola. *Mahr.*
Gandarassa, Bolam. *Cyng.*
Hoboli. *Abyssinia.*

Pereira gives a brief notice of the history of Myrrh, of which he says "the earliest "notice occurs in the Old Testament, from which it appears that this gum-resin was "an object of trade with the Eastern nations more than 3,500 years ago." In the Hebrew language it is called *Mur*, in allusion to its bitterness. The Greeks, who were well acquainted with it, called it Εμύρνα; or, in the Æolian dialect, Μύρρα. Hippocrates employed it in medicine in several diseases; and Dioscorides describes several kinds of it, the most esteemed being the *Troglodytica.* Some of the ancient poets tell us that the name of this gum resin was derived from Myrrha, the daughter of Cinyras, King of Cyprus, who fell in love with her own father, and fled to Arabia, where she was changed into a tree, which still bears her name. Notwithstanding the early knowledge of and acquaintance with the uses of myrrh, we had no accurate account of the tree which yields it until the return of Ehrenberg from his travels with Hemprich during 1820–5' in various parts of Africa and Asia. He brought with him a specimen of the tree which has been described and figured by Nees von Esenbeck under the name of *Balsamodendron Myrrha.* The first notice of the discovery of these travellers which I have met with is in Alex. Humboldt's Bericht ü die N. H. Reisen der Herren Ehrenberg und Hemprich, &c., published at Berlin in 1826. (*Pereira, Mat. Med.* ii. p. 381.)

Major Harris describes the myrrh tree as growing abundantly on the Abyssinian coast of the Red Sea to the straits of Bab-el-Mandeb, over all the barren hill sides of the low zone, inhabited by the Danakil or Adaril tribes. It is called Kurbeta, and there exist two varieties, one (producing the better description of the gum) being a dwarf shrub with deeply serrated crisp leaves of a dull green, while the other, which yields a substance more like balm than myrrh, attains a height of 10 feet, and has bright shining slightly dentated leaves. The myrrh called Hofali flows freely from any wound in the form of a milky juice possessing a perceptible acridity, which either evaporates or becomes chemically changed during the formation of the gum. The seasons for collecting it are in January, when the buds appear after the first rain, and in March, when the seeds are ripe. Every passer-by transfers such portions of it as he may find to the hollow boss of his shield, and exchanges it for a handful of tobacco with the next slavedealer whom he meets on the caravan route. The merchants also of the seacoast, before returning from Abyssinia, send into the forests that gird the western bank of the river Hawash, and bring away considerable quantities of the Hofali, which is sold at a high price. (*Piesse, Perfumery.*)

Mr. Vaughan distinctly states that myrrh is produced in Arabia, and that in the Soumali country, besides the true myrrh, a kind of which the Arabic name is *Baisa-bol*, and the Soumali, *Hebbakhade*, is obtained. In Bombay, inferior myrrh is termed *Baisabol*. (*Birdwood's Bombay Products*, p. 20.)

This gum resin, Mr. Vaughan writes, sometimes called *Murr* by the Arabs, but more commonly in this district (Aden), by the Indian name of *Heera Bol*, is collected in great quantities by the Soumalis in the north-east part of Africa, and in the neighbourhood of Hurrur, further south. It is generally brought to the large fair held at Berbera during the months of November, December, and January, and is there bought up for the most part by the Banians of India, and paid for principally in piece goods; barter, indeed, is the system upon which most commercial transactions are carried on in that district. Occasionally, the articles obtained are carried thither direct; but more frequently they are first brought over to Aden, and thence shipped for Bombay. Within the last few months small quantities of the best description of myrrh have been collected in a district 40 miles to the east of Aden, and brought hither for sale. This has been done by the Soumalis; but there can be little doubt that as soon as the native Arabs become acquainted with this valuable product of their country the trade therein will considerably increase. Four hundred and fifty hundredweights of myrrh passed through the Aden custom-house last year (1851), some of which was taken to America by American vessels, but the greater part was shipped for Bombay. The selling price here is 9½ rupees the maund of 28 lbs. (*Pharm. Journ.* xii. 227.)

From the specimens which accompanied this communication Mr. Danl. Hanbury concludes that the gum resin obtained from 40 miles east of Aden is not the myrrh of commerce, and probably is the produce of a different tree.

Under the name of *Bissa Bol* of the Arabs, *Hebbakhade* of the Soumalis, Mr. Vaughan also writes of another variety or form of myrrh which Mr. Hanbury determined from the specimens to be the kind No. 3 of Pereira, formerly called East India Myrrh, and identical with the *Myrrha Indica* of Drs. J. & E. Martiny, but quite distinct from Indian and African Bdellium. Of this Vaughan says, "This is the name of " another gum resin which is collected by the Soumalis on the opposite coast, and " brought hither for sale or exportation. In appearance it resembles the myrrh " already described, and the natives tell me that the tree from which it is obtained also " resembles the Bheera Bol tree, but is nevertheless a distinct variety. I have not met " with any description of this gum, and my impression is that the tree which produces " it is yet unknown to Europeans. It is brought over with the myrrh and other gums " by the Soumalis, but does not appear to be very plentiful, as I find that only 70 " maunds passed through the custom-house last year. Here it realizes 2½ rupees per " maund, and is sent from Aden to India and China, where it is mixed with the food " given to milch cows and buffaloes, for the purpose of increasing the quantity and im- " proving the quality of the milk. It is also used as a size, and when mixed with lime is " said to impart a bright gloss to walls which are covered with it. I am not aware to " what other purposes it may be applied, possibly it may possess other useful properties " with which the natives are unacquainted. So far as my information extends, none " of this particular gum has yet found its way to Europe." (*Pharm. Journ.* 1*st ser.* xii. p. 227.)

Mr. Johnston, also, in his travels through Adel to Abyssinia (i. p. 247), says that " there are in the country of Adel two varieties, one a low, thorny rugged looking " tree, with bright green leaves, trifoliate, and an undulating edge, is that which " has been described by Ehrenberg. This produces the finest kind of myrrh in " our shops. The other is a more leafy tree, if I may use the expression, and its " appearance reminded me exceedingly of the common hawthorn of home, having the " same largely serrated dark-green leaves growing in bunches of four or five, spring- " ing by several leaf stalks from a common centre. The flowers are small, of a light " green colour, hanging in pairs beneath the leaves, and in size and shape resemble " very much the flowers of our gooseberry tree. It belongs to Octandria Monogynia, " the eight stamens being alternately long and short, the former corresponding to the " four partial clefts in the edge of the one-leafed calyx. The fruit is a kind of berry, " that, when ripe, easily throws off the dry shell in two pieces, and the two seeds it " contains escape. The outer bark is thin transparent, and easily detached, the inner " thick, woody. When wounded a yellow turbid fluid (the gum myrrh) immediately " makes its appearance. Naturally the gum exudes from cracks in the bark of the " trunk near the root, and flows freely upon the stones immediately underneath. " Artificially, it is obtained by bruises made with stones." This plant (vide *Royle's Mat. Med.* p. 365), judging from the specimens deposited by Mr. Johnston in the

British Museum, corresponds exactly with one also in the same collection, obtained by Mr. Salt, in Abyssinia, ***Balsamodendron Kua*** of Mr. Brown's MSS., and of which Mr. Salt says he obtained from it a gum much resembling myrrh.

Pereira characterizes three kinds of myrrh, which he thus describes :—

1. *Turkey Myrrh.*—It occurs in pieces of irregular forms and of variable sizes, consisting of tears (either distinct or agglomerated) usually covered with a fine powder or dust. In a chest of this kind a few pieces of fine quality may sometimes be met with, nearly as large as a man's fist. The colour varies, being pale reddish yellow, red, or reddish brown. The pieces are fragile, semi-transparent, with a dull, in part splintery, fatty kind of fracture. In consequence of imperfect dessication the largest and finest pieces often present internally opaque whitish or yellow striœ or veins, which have been compared by Dioscorides, Pliny, and many others to the white marks on the nails. The odour of myrrh is aromatic and balsamic, peculiar, but to most persons pleasant; the taste is bitter, acrid, and aromatic. The purest, palest, and most odorous pieces are sold as picked myrrh.

2. *Myrrh in treas.*—Imported from the East Indies in chests. It consists of distinct tears or grains, which are rounded or irregular, and vary in size from that of a pin's head to a peppercorn, none of them in my specimens being so large as a small pea. They are somewhat shiny, more or less transparent, and vary in colour from pale or whitish yellow, to reddish brown. It consists of tears of myrrh intermixed with fragments of gum-arabic, and of some resin very like mastic or juniper. Many druggists in this country regard it as merely the siftings of the finest kind, but I cannot agree with them in this opinion.

3. *East India Myrrh.*—Formerly this was the only kind imported from the East Indies. It occurs in pieces which are darker coloured than those of the so called Turkey Myrrh, and whose average size does not exceed that of a walnut. It is often mixed with other substances, particularly with Indian Bdellium and with a substance of similar appearance to dark red coloured Senegal gum.

Myrrh, according to Ehrenberg, exudes like cherry tree gum from the bark of the tree. It is at first soft, oily, and of a pale yellow colour, but by drying becomes darker and redder. It is imported from the East Indies in chests, each containing 2 to 4 cwt. Formerly the finest kind was brought from Turkey, and an inferior sort from the East Indies, but at the present time nearly the whole is brought from India. Sometimes the same chest contains myrrh of all qualities, which is then termed myrrh in sorts, but commonly it is brought over more or less sorted. (*Pereira, Mat. Med.* ii. 381.)

Myrrh is only partially soluble in water, alcohol, or ether; the first of those liquids takes up the gum, principally the two latter, the resin and the oil. Water takes up more of the myrrh than alcohol does. Alkaline solutions are good solvents for myrrh. A few drops of nitric acid dropped on a small fragment of myrrh or on a concentrated tincture develop a red colour.

Myrrh was analysed in 1816 by Pelletier and in 1819 by Braconnot and by Brandes.

—	Brandes.	Braconnot.	Pelletier.
Volatile oil	2·60	2·5 } 23·0	34
Resin { Soft	22·24 }		
Resin { Hard	5·56 }		
Gum { Soluble	54·38	46·0 }	66
Gum { Insoluble	9·32	12·0 }	
Salts (benzoates, malates, phosphates, sulphates, and acetates of potash and lime).	1·36	—	—
Impurities	1·60	—	—
Loss	2·94	16·5	—
Total	100·00	100·00	100

East Aden Myrrh.

The only information regarding this myrrh is that contained in Mr. James Vaughan's communication, already cited, and in Mr. Daniel Hanbury's notes thereon. Mr. Vaughan says " within the last few months small quantities of the best description " of myrrh have been collected in a district forty miles to the east of Aden, and "brought hither for sale. This has been done by the Soumalis; but there can be "little doubt that as soon as the native Arabs become acquainted with this valuable "product of their county, the trade therein will considerably increase." Of the specimen which accompanied these remarks Mr. Hanbury writes, "the myrrh which "is produced forty miles to the eastward of Aden is a gum resin distinctly different, "and, as the author suggests in another communication, doubtless the produce of "some other tree than that affording common myrrh. It is in irregular pieces, "varying in size from that of a walnut to less than that of a pea, not coated with "dust like pieces of Turkey Myrrh, but having a somewhat shining exterior. Each "larger piece appears to be formed by the cohesion of a number of small rounded "somewhat transparent externally shining tears or drops. The fracture much re- "sembles that of common myrrh, but wants the semi-circular whitish markings. "In odour and taste it agrees closely with true myrrh. Portions of a semi- "transparent brown papyraceous bark are occasionally attached to pieces of it. I "have recently noticed a small quantity of this myrrh in the hands of a London "drug broker." (*Pharm. Journ.* xii. p. 227.)

Imports of Myrrh into the Bombay Presidency.

—	1869–70.	1870–1.	1871–2.	1872–3.
Bysabole:	cwt.	cwt.	cwt.	cwt.
From Red Sea	82	—	—	—
Aden	212	252	303	224
Africa	—	—	3	—
Heerabole:				
From Africa, coast	—	40	—	16
Red Sea	117	105	198	188
Aden	727	188	478	290
Persian Gulf	1	—	15	—
Total	—	—	—	—

Exports of Myrrh from the Bombay Presidency.

—	1869–70.	1870–1.	1871–2.	1872–3.
Bysabole:	cwt.	cwt.	cwt.	cwt.
To China	273	—	59	138
Heerabole:				
To United Kingdom	394	483	196	493
Suez	—	—	1	—
Turkey	—	—	—	6
Persian Gulf	—	—	2	13
China	—	—	6	29
Sonmeanee and Mech	—	—	1	—
Laccadive and Maldives	1	—	—	—
Total	—	—	—	—
To Madras Presidency	—	—	—	5

Bibliography of Myrrh.

Pharmacopœia of India, p. 51.
Pereira, Materia Medica, vol. ii. p. 381.
O'Shaughnessy, Bengal Dispensatory, p. 285.
Royle, Materia Medica, 4th ed. p. 365.
United States Dispensatory, pp. 557.
Pharmaceutical Journal, xii. 226. 1859, p. 217.
Carson, Illustr. of Medical Botany, i. 28, pl. 20.
Johnston, Travels through Adel to Abyssinia, i. p. 247.
Flückiger and Hanbury, Pharmacographia, p. 124. London, 1874.
Guibourt, Hist. Nat. des Drogues, iii. 469.

Balsamodendron Roxburghii, *Arn.*

Amyris commiphora. *Roxb. Fl. Ind.* ii. 244.
Amyris agallocha. *Roxb. Hort. Beng.*
Balsamodendron agallocha. *Wi. Ill.* i. 185. *Drury Use, Pl.* 62.

Tree, trunk crooked, and clothed with many drooping crooked branches down to the ground; branchlets often ending in thorny points; calyx 4-toothed; petals, four leaves alternate, petroled, oval or elliptic, serrulate, smooth on both sides; at the base or apex of the petiole on each side is generally a small leaflet giving the appearance of a ternate leaf; flowers on short peduncles axillary, small red, aggregate on the buds by the former year's leaves; berry drupaceous, red, smooth, size of a currant; 2-celled, and 1-seeded.

Hab.—Silhet. Assam.

Sir Wm. Hooker was clearly of opinion that this is a distinct tree from that which yields Googul in Sinde. Alluding to Dr. Stocks' *Balsamodendron Roxburghii* he writes " in the MS. this gentleman had called the plant by the name of *Balsamodendron* " *Roxburghii*, from an idea that it was identical with the *Balsamodendron Roxburghii* " of Arnott (the *Amyris commiphora* of Roxburgh), a native of north-eastern Bengal. " A comparison of specimens of both in my herbarium has satisfied Dr. Wallich and " myself that the two species are very distinct. It is true it bears the same name " among the Bengalese as the Scinde plant, but that name appears to be given to " different plants yielding a somewhat analogous product, as for example, the Googul " or Googil of the Coromandel coast, which Dr. Stocks observes is the *Boswellia* " *glabra*. But it is extremely improbable that a plant of the North-eastern frontier " of Bengal should be identical with one of Scinde, where the vegetation bears a " striking resemblance to that of Syria or Arabia. Dr. Roxburgh too, observes that " though his plant, when broken or bruised, diffuses a grateful fragrance, like that " of the finest myrrh, yet that the juice never congeals, but is carried off by evapora- " tion, leaving little or nothing behind, and all that he could ever procure was a " very minute portion of gummy matter, which certainly resembles myrrh both in " smell and appearance, but has no tendency to be even tenaceous or elastic." The excellent Dr. Royle, however, rather inclines to the opinion that this tree, when old, does yield a gum resin, closely resembling myrrh, because that which he examined was said to come from the hills at the foot of which the tree is found. Be that as it may, it is very certain that the Mukul or Googul Balsam tree of the Persian Gulf described by Dr. Stocks is a very distinct species; and the gum resin it yields is much more likely to be that of the ancient writers on the subject, for it is assuredly the genuine Googul of the Bazaars of Hydrabad and Kurrachee, and that which is exported from Bombay. (*Hook. Journ.* i. 257. 1849.)

In a communication from Mr. John Lovell some information is given relative to this product; he writes—I forwarded from Sholapoor to Dr. Cleghorn, then at Madras, specimens of the fluid gum obtained from the *Amyris commiphora*, collected at Tooljapoor, also flowering specimens of the plant and cuttings for propagation in the Horticultural Society's garden at Madras, and concerning the singular hardy nature of the said plants or trees I may add that I had several cuttings as thick as, and some thicker than short walking sticks which flourished after severance by knife from the parent stock in a very astonishing manner putting forth buds and without any symptom of withering or dying standing upright on the chunam floor in a bath or bedroom. Thus they continued for many weeks during which period I frequently inspected them both at Sholapoor and Bellary, to which latter station I conveyed them from the former and at length planted them in my garden at Bellany. From Dr. Cleghorn I had the pleasure to receive amongst others the following remarks:—" A careful drawing " has been made of the flowering specimens, the plant is evidently a true Balsam- " odendron, the cuttings which you kindly sent me are thriving very well in the " Horticultural Gardens but they have not as yet flowered. Since my return from " Ireland I have had access to my boxes and found some of the residuum which " remained after pouring off the more fluid Balsam in India, and subsequently sun- " dried. The exuded juice from the plants or trees, produced from scarification of the " bark when fresh and for some time subsequent was found in bottles milky or turbid " in appearance, and most fragrant, but gradually became finer and clearer by age and " exposure to the sun as the specimen already forwarded to your museum from Ireland " demonstrates, and now, I have the pleasure to add the promised residuum. The pro- " portion which in quantity one bears to the other I have no means of determining, " and I fancy of little importance if I had, as analysis will doubtless settle the point "and certainly more to the purpose. I shall at any future time be pleased to receive "from you or Dr. Waring any account of experiments made with the gum tending to "its utilization or employment in medicine. The gum might be procured in larger "quantities, and the tree easily propagated. The modern Hindoos or Brahmins about "Tooljapoor and Mandu (ancient capital of Malwa under Mahomedan rule) knew "nothing of its uses, or for what purpose employed (beyond burning a few of the "dried sticks to perfume their shrines, it may be, though not often done) in ancient "times, whence it came, or by whom planted, where found, and only in those places "have I ever seen the tree from Cape Comorin to the latitude of Mandu in Malwa "on the Vindian range of mountains, parallel, or a little north of Calcutta, and my "station Mhow. A pretty wide and general traverser of the countries lying between "both in length and breadth of our Madras and other Presidencies have I been, few "men of our service have seen so much of the portions of India indicated."

Balsamodendron Mukul, *Hook.*

Hooker Journ. (1849) i. 257.

Balsamodendron Roxburghii. *Stocks, Bombay Trans.* 1847.

Small spiny tree 4-6 feet, young leaves clad with glandular hairs, leaves simple and obovate, serrate-dentate anteriorly or trifoliate, middle leaflet rhomboid or oval acute, lateral sometimes minute, and entire, generally serrate, half the size of the terminal leaflet; calyx campanulate, petals ligulate, reflexed at the apex; stamens 8-10 the four opposite the petals shortest. Fruit ovate acuminate, cleft to the base into two valves, nut involved in a 4-cleft yellow pulp, the limbs of which meet at the apex.

Hab.—Scinde, on rocky ground.

The following synonyms for the resin are given by Baboo Moodeen Shereef

Moql, Moqle-arzag, Aflatan. *Ar.*	Gugul. *B.*
Boe-jahudan. P.	Kou-shikaba. *Sans.*
Gogil. *H.*	Guggula. *Mahr.*
Gugal. *Duk.*	Gugal. *Guz.*
Mai-sakshi. Mai-sakshi. *Tel.*	Gugula, Jatayu or Javayu Rata dummula. *Cyng.*
Guggala. *Can.*	

This shrub is called Googul or Guggur by the hill Beloochees, who indeed do not know it by the name of Mukul. It yields the gum-resin googul, which they collect, and bring to the bazaars of Hydrabad and Kurrachee, where it sells at the rate of two rupees the maund of 80 lbs. At Bombay its tariff valuation is two rupees the maund. It is collected in the cold season by making incisions with a knife in the tree, and letting the resin fall to the ground. Hence the dirty and impure state in which it is found in the shops. I have obtained it from September to February, and have found it exude in large tears from a clean incision, of the colour, consistence, and opacity of pus laudabile. My informants say that from half to a whole seer is yielded by a single tree. It is esteemed cordial and stimulant. It hardens and turns brownish black very slowly. Made up into a cake with bajree flour, it is commonly given to horses and cattle when they have a cold. The dealers from Cabool have a custom of administering it to their horses in the cold season, thinking that it keeps them in health and condition. The fruit and young shoots are applied to a similar purpose. The gum is made into a plaster and used to discuss tumours and boils, and is regarded as efficacious in expelling the guineaworm, both taken internally and applied to the tumour. It is extensively employed by the Hindoos as incense for burning in their temples, although its smell is by no means agreeable. It is also much prized by builders, who mix it with the mortar and plaster used in the construction of houses of a somewhat superior description where durability is an object. The googul is boiled in water for a considerable time, when its spirit (as they phrase it) is communicated to the water, and the dregs are thrown away. This solution of the gummy part, which according to Newman's analysis should be six drachms two scruples in every ounce, is mixed with the lime, and employed with confidence to make the plaster adhere strongly, and to prevent it from crumbling and splitting. The googul water is sometimes washed over the walls by itself. (*Hook. Journ.* 1849, 1. 2.)

In addition to the above particulars Dr. Stocks secured a translation of the article "Mukul" in the Makhzan-el-Adwiyed, *i.e.*—

Mukl. In Puryani ***Mukla***, in Rumi ***Budliyun***, in Arabic ***Kafr*** (bitumen or pitch, also called ***Kafr ul Yahud*** or Jew's pitch) and ***Kawar***, in Persian ***Bui Tahadan*** (the Jew's perfume because that people use it in fumigation), in Hindustani ***Gugal***. It is the gum of a large tree about the size of the Kundur (Olibanum) growing plentifully upon the shores of the sea of Uman (the sea on the east coast of Arabia) and in Sanjar (Khorasan ?) and India. Its general characteristic is bitterness, and it is of many kinds, as *e.g.* 1st, the *Mukl-i-arzah* (bluish Bdellium) in colour reddish and bitter, 2nd, *Mukl-el-Yahud* (Jew's Bdellium) of a yellowish tinge; 3rd, *Mukl-i-Sakalbi* which is clouded, impure, black, and soft; 4th, *Mukl-i-Arabia* (Arabian Bdellium) which grows in Yemen, and is of the colour of the *Badanjan* (ripe fruit of the egg-plant, *i.e.* greenish black). The best kind is clear, pure, and brilliant, viscous, adhesive, soft, sweet smelling, yellow, and bitterish. When thrown upon fire it emits an odour like the laurel, and readily dissolves in water. It must be unmixed with wood, straw, sand, earth, or such matters. Its properties last for twenty years, when old its bitterness increases, and the older it is the darker it becomes, exchanging its softness for dryness and hardness, especially the Arabic, as they mix it with myrrh." (*Hook. Journ., ib.*)

To this translation Dr. Stocks adds that his friend Assistant Surgeon Carter showed him specimens of the Mukul gum collected by him on the southern coast of Arabia, together with numerous other gums, all accompanied by admirable drawings of the trees producing them. There is therefore some error in the statement of Dr. Malcolmson that Bdellium is not produced in Arabia. Moreover the *Mukul*, and the tree producing it, are from Dr. Carter's specimens identical with the Scinde Googul, and its tree, as might be expected from the great similarity between the vegetation of the rocky part of Scinde and that of Arabia. The range of our Googul tree is therefore Arabia (Dr. Carter), and according to my own observations in rocky ground throughout Scinde, at Deesa in Marwar, and lately in Beloochistan proper. It flowers in March and April and the leaves and young shoots appear in April and May. In sheltered situations, as under the bank of a watercourse, it may be found in fruit, flower, and leaf for the greater part of the year. (*Hook. Journ.* 1849, i. p. 257.)

In his catologue of Bombay products, Dr. G. M. Birdwood has sketched exhaustively the ancient history and associations of this drug. He observes of Googul that it is "generally thought to be the Bdellium (Bdolach) of Gen. ii. v. 12, and Numb. " xi. v. 17, and the βδέλλιον and μαδελκὸν vel μαδαλκον of Dioscorides." Lassen however has suggested that Bdolach means musk, and not Bdellium, and it is difficult to resist his conclusion based as it is not only on the description given of Bdolach in Numbers, but also on its affinity to the Sanscrit word Madalaka, which is thought to mean musk. In Genesis the word occurs in the passage "there (that is in " the land of Havilath, compassed by the river Pishon) is Bdellium and the onyx " stone." The Pishon being considered by commentators to be the Indus, the text seems so far to support the view of Googul being the ancient Bdellium or Madalcon. But the musk region of the Himalayas may be as appropriately described as compassed by the Indus, as Sindh and the Punjab. The association of Bdolach with gold points also to Thibet. Nothing is known of the locality of Hawilath independently of the Pishon. So much for the positive argument for Bdolach being musk. With regard to the negative, neither the descriptions of the Bible, Dioscorides, nor Pliny in the least resemble *Googul*, while all indicate musk. Pliny gives Bactriana as the country of Bdellium, but says it also comes from Arabia and Media, the Median being called " peraticum" (πέρατα γῆς) or "from the uttermost parts of the earth." The musk deer is not only found in the Himalayas but in Siberia, Tonquin, and Cochin China, and a substance analogous to musk (Hyraceum ?) is brought to Bombay by Zanzibar merchants. *Googul*, however, is found not only in Northern India, but in Arabia; nevertheless I am of opinion that Pliny never meant *Googul* by Bdellium, and that probably his "Indian Myrrh" and "Scordastum" refer to the modern Bdellium of India. It is strange that although familiar with Castoreum, no ancient writers mention musk unequivocally; Ætius (A.D. 550) being the first who describes it. The etymology of musk (μόσχος) is not determined. The connection of the "onyx stone" with "bdellium" in Genesis (cap. ii. v. 11), renders it necessary to remark that the word "onyx" is used in another sense in Scripture, according to Calmet than that of the stone *Shohem*. Thus the word *Shecheleth* is translated by the seventy as "onyx" (ονυξ a nail), meaning the celebrated "odoriferous shell" of the ancients; although others understand by it *Ladanum* (the balsam of *Cistus creticus*, W.; *C. ladaniferus*, W.; &c.) and Bdellium. Pliny says of Bactrian Bdellium that it "is shining and dry, and covered with numerous white spots resembling the finger " nails." And such Bdellium would appear to have been the βδέλλη ὄνυξ of Democritus, an obscure medical writer quoted by Saracenus in his *Scholia in Dioscoridis* and of Galen, as quoted by Salmasius in his *Plinianæ Exercitationes.* Salmasius states that from the Greek words μαδελκον, μαλχὴ, the arab *Molochil* (Mukul) is derived; which if true would lessen the force of Lassen's arguments in favour of the Bdellium of the Bible being musk, if they were etymological only. Bochart asserts " that the Bdolach of the Bible is neither a stone nor Bdellium but a shell, genus " *Unio*."

Googul or Indian Bdellium is sometimes sold for and considered as an inferior kind of Myrrh. It is in roundish pieces of a dull dark red colour more moist than Myrrh, and not brittle like it; softening even with the heat of the hand, bitter, and a little acrid in taste, with a less agreeable odour. It often has portions of the birch-like bark adhering to it. (*Royle, Mat. Med.* p. 366.)

Balsamodendron pubescens, *Stocks.*

Stocks in Hook. Jour. Bot. 1849, i. p. 257. *Boiss. Fl. Orient.* vol. ii. p. 3.

Small tree, unarmed, young shoots clad with a short furfuraceous pubescence; leaves bi-trijugate; leaflets, obovate, entire, lateral, often oblong-obovate; calyx, urceolate, campanulate; petals obovate, acute, apices patent; stamens, equal in length; disc equally toothed; ovary bicelled; fruit globose acute; valves two cleft half way; pulp, orange coloured, 4-toothed upwards; apex of nut naked; nut ovate, obtuse.

Hab. Beloochistan, and the hills which separate that province from Scinde, probably also of Affghanistan, attaining its southern limit about Kurrachee.

This shrub is called Bayee by the Hill Belooches, who make no use of it. Its young shoots and buds are remarkably fragrant when bruised. In the cold season it yields a small quantity of a tasteless inodorous brittle gum almost entirely soluble in water. It flowers in March and April, and its leaves and young shoots appear in April and May.

IMPORTS of GUM BDELLIUM into the BOMBAY PRESIDENCY.

—	1869–70.	1870–1.	1871–2.	1872–3.
From	cwt.	cwt.	cwt.	cwt.
Red Sea - - -	8	139	101	200
Aden - - -	—	—	—	7
Persian Gulf - -	—	63	—	70
Soumeanee and Meckrun -	—	43	—	—
Total - -	—	245	—	277
Scinde - - -	—	—	—	426
Total - -	—	—	—	703

Bibliography of Googul.

Ulfaz Udwiyeh, No. 864, 968, 1383, 222.
Taleef Shereef, No. 809, 944.
Forskahl, Materia Medica, p. 158.
Arrian, Expeditio Alexandri, lib. vi. c. xxii.
Stocks in Hooker Journ. Bot. 1849, i. p. 257.
Pharmacopœia of India, p. 53.
Agra Exhibition Catalogue, 1867.
Ainslie, Materia Medica, i. p. 69. (1826.)
Smith's Chinese Materia Medica. 1871.
Lindley's Flora Medica, p. 170.
O'Shaughnessy's Bengal Dispensatory, p. 287.
Birdwood's Bombay Products, p. 262.
Royle, Materia Medica, 4th ed. p. 366.

Mangifera Indica, *L.*

Roxb. Fl. Ind. i. 641.

Tree, leaves alternate, lanceolate, acuminated, glabrous; calyx 5 cleft; petals 5; panicles terminal, much branched, pubescent, erect. Drupe obliquely oblong, or somewhat reniform; seed solitary, flowers small, greenish yellow.

Hab.—Common everywhere in India.

Am. *H.*
Racha-mamidi, Tiyya-mamidi, Mamidi, Mamari, Mavi, Mamidi-chettu, Ela-mavi, Gujju-mamidi, Mamadi-chettoo. *Tel.*
Maa, Ma-maram, Ma-marum, Maah-marum. *Tam.*
Tha-yat, That-yat, Tha-rat, That-rat. *Burm.*
Am. *B.*
Mavena. *Can.*
Palam. *Jav.*
Kapalam. *Lampung.*
Ampalam, Mampalam. *My.*
Mava, Mangai, Mangoe. *Mal.*
Makandamu. *Sans.*
Ambo. *Uriya.*
Aetamba-gaha. *Cyn.*
Am, Aum. *Ass.*

In this instance, as in some others, confusion of different substances under the name of Mango gum is by no means uncommon, as for instance, apparently so in the Punjab Products (p. 416), where a specimen is recorded from Delhi of a white gum from the Mango tree. The best description of the true Mango resin as we understand it is contained in Drury's useful plants (2nd ed. p. 287), where it is stated that "from "wounds in the bark issues a soft reddish-brown gum resin, hardening by age and "much resembling bdellium. Burnt in the flame of a candle it emits a smell like "that of cashew nuts when roasting. It softens in the mouth and adheres to the "teeth, and in taste is somewhat pungent and bitter. It dissolves entirely in spirit "and partly so in water."

In collections the gum of Anacardiaceous trees, especially of *Anacardium occidentale* is confounded with Mango gum, although so entirely different from it in character, and what appears to be the gum of *Moringa pterygosperma* has been forwarded to England as Mango-gum. A sample of veritable Mango-gum with a strong persistent odour as of ripe Mangoes was sent from Aden through Bombay to the Vienna Exhibition as *Hing*, and casually it had the appearance of some varieties of Asafœtida, but the odour is very distinct. Another sample under the name of *Habk-hadie* was received through the same channel and is scarcely different. The name of *Hebbakhade* seems to be employed by the Soumalis for a very distinct gum resin, as cited by Vaughan, who gives it as a synonym for the *Bissa Bol* an inferior kind of myrrh collected by the Soumalis. The sample of Hebbakhade sent by Mr. Vaughan was examined by Mr. D. Hanbury and determined as the myrrh formerly known as East India Myrrh.

The specimen of Mango-resin received under the name of Habk-hadie is of a dark, rather warm brown, mottled with opaque pale ochraceous irregular spots, patches, and streaks, so that it is not unlike some samples of Asafœtida, or indeed of Benzoin of an amygdaloid kind, only that it is darker in colour, and the pale tears and streaks are less in quantity than the connecting dark brown resin. The fracture is shining but not smooth or conchoidal. To the touch it is similar to Benzoin in hardness, and in friability, and is not more adhesive to the fingers. Probably it acquires firmness readily by age. There is a peculiar pungency in the taste of a rather aromatic kind, leaving a distinct bitterness. In odour it resembles nothing so much as ripe mangoes, and this odour seems to be strong and persistent.

The remark is appended in the Madras Catalogue of 1855 to Mango gum exhibited by Dr. Cleghorn to the effect that it is sold as "Googlut" at Sholapore.

Sub. Div. β.—BENZOIN KIND.

This sub-division includes such gum resins as are characterized by the presence of Benzoic acid, or which may be employed as incense, as Olibanum and similar substances. The two typical gum resins of the group being Gum Benzoin and Gum Olibanum, these may be held to represent the two similar but allied kinds of fragrant gum resins associated under this sub-section. Formerly it was customary to class as balsams all the resins in which Benzoic Acid was present, whether solid as in the case of Benzoin, or not. This method has of late been abandoned, and the limitation given to the term *Balsam* will be defined under that sub-division.

Styrax Benzoin, *Dryand.*

Phil. Trans. 77, p. 308 t. 12. *Steph. & Church. Med. Bot.* ii. t. 112. *DCand. Prod.* viii. 260.

A tree of considerable size, but small when tapped, of quick growth; branches and footstalks round, downy; leaves oblong, acuminated, smooth above, white and tomentose underneath; acemes compound, axillary, nearly as long as the leaves, flowers on one side with short pedicels; calyx campanulate, obscurely 5-toothed; corol of 5 petals of a greyish white colour, which are perhaps united at the base, four times longer than the calyx; stamens 10 inserted into the receptacle; ovary superior, ovate, tomentose; style filiform; stigma simple. (*Esenb. & Eberm.* 111.)

Hab.—Sumatra. Java. Borneo. Siam, &c.

The native names of Benzoin are—

Looban, Ood. *H.*
Sambranee, Malacca-sambranie. *Tam.*
Ood, Loobanee-ood. *Dec.*
Cominjan, Kamenian. *My.*
Lawban, Hékkoo-kama. *Burm.*
Liban. *A.*
Hussee-ool-jawee, Hussee-looban. *P.*

The mode of collecting Benzoin in Sumatra is thus summarized by Pereira:—"When "the tree is six years old longitudinal or somewhat oblique incisions are made in the "bark of the stem, at the origin of the principal lower branches. A liquid exudes, "which, by exposure to the sun and air soon concretes, and the solid mass is then "separated by means of a knife or chisel. Each tree yields about three pounds of "Benzoin annually, for the space of ten or twelve years. That which exudes during "the first three years is white and is denominated *head benzoin.* The benzoin which "subsequently flows is of a brownish colour, and is termed *belly benzoin.* After "the tree is cut down the stem is split, and some benzoin scraped from the wood, "but its colour is dark, and its quality bad, owing to the intermixture of parings of "wood and other impurities; this sort is called *foot benzoin.* The relative values of "head, belly, and foot benzoin are as 105, 45, 18. Benzoin is brought down from the "country in large cakes (called by the natives *tampangs*) covered with mats. In "order to pack it in chests these cakes are softened by heat, the finer by exposure to "the sun, the coarser by means of boiling water." (*Mat. Med.* ii. 683.)

Pereira classifies Benzoin under the heads Siam Benzoin and Sumatra Benzoin.

Siam Benzoin.—Crawford says that the Benzoin of Siam is procured from Lao. He also states that a substance resembling and hitherto confounded with benzoin, produced in Lao, Raheng, Chiang-mai, and La Kou, is abundantly found in Siam. The tree producing it cannot be, he thinks, the *Styrax Benzoin,* as it grows so far north as the twentieth degree of latitude.

Siam Benzoin is brought to England, either direct from Siam, or indirectly by way of Singapore. It includes the best commercial sorts, or those known in commerce as Benzoin of the finest quality. It occurs in tears in irregular lumps and in cubical blocks, but unlike the Sumatra sort, it never comes over enveloped in calico. It is in general distinguished from the other sorts by its warmer or richer (yellow, reddish, or brown) tints. The dealers distinguish five or six varieties, the three best sorts being included under the name of Yellow Siam Benzoin, and the two or three inferior kinds being called red or brown Siam Benzoin; the designations yellow, red, or brown being used on account of the tint of this resin by which the tears are agglutinated; but the division is altogether arbitrary, the colours passing universally into one another. I shall therefore adopt another arrangement.

α. Siam Benzoin in tears (Yellow Benzoin in the tear). This kind seems to be identical with "the true benzoin in tears" which Savary says was brought in considerable quantity to Paris by the attendants of the Siamese Ambassadors. It consists of irregular flattened pieces, some of which are angular, and the largest of them barely exceeding an inch in length. Externally these pieces are shiny or dusty from their mutual friction, and are of an amber or reddish yellow colour; they are brittle, and may be easily rubbed to powder. Internally they are translucent or milky, and frequently striped; they have a pleasant odour, but little or no taste. There is an inferior sort of Siam benzoin in tears, which consists of loose drops mixed with pieces of wood and other impurities. It is worth only one-fourth of the price of clean good-sized tears or clean lump.

β. Siam or lump Benzoin.—The finest kind consists of agglutinated tears (white or yellow lump benzoin); more commonly we find the tears are connected by a brown resiniform mass, which, when broken, presents an amygdaloid appearance, from the white tears imbedded in the mass (amygdaloid benzoin).

Inferior sorts of lump benzoin are reddish (red lump benzoin).

γ. Translucent Benzoin.—From my friend Dr. Royle I received a sample of Siam Benzoin, the properties of which are somewhat different from the preceding. The small masses consist of agglomerated tears, which instead of being white and opaque, are translucent, or, in a few instances, almost transparent.

Sumatra Benzoin.—Though placed here second this sort is the more important, being in many countries the only kind known. It is rarely imported direct from Sumatra, but in general indirectly by way of Singapore or Bombay, and now and then from Calcutta.

It occurs in large rectangular blocks, marked with the impression of a mat, and covered with white cotton cloth. When broken, we observe but few large white tears in it. The mass is generally made up of a brown resiniform matter with numerous white, small pieces or chips intermixed, which thereby give the broken surface a speckled appearance, somewhat like that of a fine-grained granite.

The qualities of Sumatra Benzoin are distinguished as firsts, seconds, and thirds.

α. The first Sumatra sort occurs very seldom and only by single chests, for which 50*l*. or more the cwt. are paid in the Russian market.

β. The second Sumatra sort is also marbly but not so white, and is also mostly taken for the Russian market at 20*l*. to 30*l*. the cwt. Thirty chests of this sort perhaps are seen before one of the first quality is met with.

γ. The third Sumatra sort is browner, and less or not at all marbly. It fetches from 5*l*. to 10*l*. per cwt., and forms the usual commercial quality (common or brown benzoin). five times as much of this quality is met with as of all the other sorts put together.

There is a very inferior sort of benzoin (inferior Bombay Benzoin) which invariably comes by way of Bombay. If it be the produce of Sumatra it is remarkable that it never comes by way of Singapore. (*Pereira, Mat. Med*: ii. p. 684.)

Benzoin has a fragrant odour, with very little taste, but when chewed for some time leaves a sense of irritation in the mouth and fauces. It breaks with a resinous fracture, and presents a mottled surface of white and brown or reddish brown; the white spots being smooth and shining, and even translucent, is usually more or less rough and porous, and often exhibits impurities. In the inferior kinds the white spots are few or entirely wanting. Benzoin is easily pulverized and in the process of being powdered is apt to excite sneezing. In sp. gr. is 1·063 to 1·092; when heated it melts and emits thick, white, pungent fumes which excite cough when inhaled, and consist chiefly of benzoic acid. It is wholly soluble, with the exception of impurities, in alcohol, and is precipitated by water from the solution, rendering the liquor milky; it imparts to boiling water a notable proportion of benzoic acid. Lime water and the alkaline solutions partially dissolve it, forming benzoates from which the acid may be precipitated by the addition of other acids. Its chief constituents are resin and benzoic acid, and it therefore belongs to the balsams. The white tears and the brownish connecting medium are said by Stolze to contain nearly the same proportion of acid, which according to Bucholz is 12·5 per cent. and to Stolze 19·8 per cent. In a more recent examination by Kopp, the white tears were found to contain from 8 to 10 per cent. of acid and the brown 15 per cent. The resin is of three kinds, one extracted from the balsam with the benzoic acid by a boiling solution of carbonate of potassa in excess, another dissolved by ether from the residue, and the third affected by neither of these solvents. Besides benzoic acid and resin the balsam contains a minute proportion of extractive, and traces of volatile oil. Benzoin is said to retard the oxidation of fatty matters and thus to prevent rancidity. (*U.S. Dispensatory*, pp. 165.)

Benzoin has been the repeated subject of chemical analysis. It was analyzed in 1811 by Bucholz, in 1816 by John, in 1823 by Stoltze, and in 1845 by Kopp. (*Pereira, Mat. Med.* ii. 685.)

—	Bucholz.	John.	Stoltze.		
			White.	Amygdaloid.	Brown.
Volatile oil	—	—	traces	traces	traces
Benzoic Acid	12·5	12·0	19·80	19·42	19·70
Resin {Yellow	83·3	84·5	79·83	27·10	8·80
Resin {Brown			0·25	50·53	69·73
Matter like Balsam of Peru	1·7	—	—	—	—
Aromatic extractive	0·5	0·5	—	0·25	0·15
Woody matter and other impurities	2·0	2·0	—	2·60	1·45
Water and loss	—	0·25	0·12	0·10	0·17
Salts (benzoates and phosphates)	—	0·75	—	—	—
	100·	100·	100·	100·	100·

In the Museum Collection is a specimen of white benzoin in a flattened cylinder, wrapped in palm leaf from Malacca which is very fine and peculiar; specimens of Benzoin in the collection are from Penang, Calcutta, Sumatra, and Siam.

Terminalia angustifolia, *Wi. & Ar.*

Wi. & Arn. Prod. i. 312. *Jacq. Hort. Vind.* iii. t. 100.
Terminalia Benzoin, *Linn. f.*
Catappa Benzoin, *Gærtn.* ii. t. 127.
Croton Benzoe, *Linn. Mant.* 297.

Tree 30–40 feet; calyx campanulate, 5–cleft; petals none; leaves alternate linear-lanceolate, attenuated at both ends, crowded at the ends of the branches; under side and petioles pubescent or hairy; drupe compressed, 2 winged gibbous on one side; stamens in two rows; seed almond-like, flowers spiked, small, green odoriferous.

Hab.—East Indies.

This tree produces a gum-resin in some respects resembling Benzoin, which is procured by wounding the tree and "is composed," says Dr. Royle, "of large white and "light brown pieces easily broken between the hands. When gently dried it forms a "white powder formerly in great request as a cosmetic; it has a most agreeable scent; "but the most striking ingredient of this resin is the Benzoic Acid; in the churches "in Mauritius this resin is used as incense."

If the specimen contained in the Museum is authentic, it consists of large irregular shaped pieces often several inches in diameter, externally dull, when broken of a shining resinoid appearance with a vitreous fracture, of a more or less vinous brown colour, brittle, easily reduced to a dirty white powder, and wherever struck or bruised in the same manner powdery. Some fragments are homogenous, but usually there is a more or less granitic structure as in common Benzoin; in odour it faintly resembles Benzoin, but more like that of Olibanum, and like the latter is rendered somewhat sticky by the warmth of the hand.

IMPORTS of GUM BENZOIN into the PRESIDENCY of BOMBAY.

—	1869–70.	1870–1.	1871–2.	1872–3.
From	cwt.	cwt.	cwt.	cwt.
Aden	28	—	—	—
Persian Gulf	—	—	5	—
Straits Settlements	4,171	5,286	5,834	4,902
China	27	82	33	—
Siam	146	—	8	—
From				
Bengal Presidency	442	56	94	56
Madras Presidency	—	—	1	89
Total	—	—	—	—

EXPORTS of GUM BENZOIN from BOMBAY PRESIDENCY.

—	1869–70.	1870–1.	1871–2.	1872–3.
	cwt.	cwt.	cwt.	cwt.
United Kingdom	245	7	385	1,093
Suez	—	—	4	—
Beyrout	—	—	—	2
France	—	—	22	—
Turkey	—	—	2	8
Mauritius	—	1	—	1
Africa, coast	63	18	82	134
Red Sea	21	48	107	26
Aden	16	5	7	37
Persian Gulf	78	100	177	198
Sonmeanee and Meckran	8	9	9	6
Laccadive and Maldive Islds.	—	1	2	3
Straits Settlements	2	—	28	2
Total	433	189	825	1,510
To				
Madras Presidency	190	110	156	267
Scinde	121	56	62	45
	—	—	—	—

Bibliography of Benzoin.

Pereira, Materia Medica, ii. p. 682.
United States Dispensatory, p. 165.
Pharmaceutical Journal, Aug. 1863, vol. v. p. 77.
Journal de Pharmacie, 3rd ser., iv. 46.
Annal. der Chem. und Pharm., cxix. 136.
Marsden, History of Sumatra, 3rd ed., p. 134.
Crawford, Hist. Ind. Archipelago, i. 518, iii. 418.
Crawford, Journal of an Embassy to Siam, p. 407 (1828).
Royle, Materia Medica, 4th ed., p. 512.
Guibourt, Hist. Nat. des Drogues, ii. 550.
Flückiger and Hanbury, Pharmacographia, p. 361.

Olibanum.

The exhaustive memoir by Dr. Birdwood of the history and associations of this product leaves nothing to be desired except that it should be more extensively known; with this view it is given *in extenso* as an appendix to this report.

The native synonyms of Olibanum as given in the Supplement to the Pharmacopœia of India are—

Bastaj, Kundur, Luban. *Ar.*
Kundur. *P.*
Sel-gond Kundur. *H.*
Kundur. *Dec.*
Parangi-shambirani, Kundurukkam-pishin. *Tam.*
Parangi-sambrani, Anduga-pisunu. *Tel.*
Vella-kundirukkam. *Mal.*
Kundro, Salai, Salai-gun. *B.*
Salasi-niryasam. *Sans.*
Kundrikam. *Cing.*

The species determined by Dr. Birdwood as Olibanum-producing trees are those hereafter named.

Boswellia Carteri, *Birdw.*

Birdwood in Linn. Trans., xxvii. p. 143, t. 29, 30. *Carter in Journ. Bom. Br. Roy. As. Soc.*, ii. (1847).

Small tree; branches pubescent or tomentose at the extremities, leaflets 7–10 pair ovate-oblong, undulate or crenate-undulate, broadly rounded, or truncate at the base; racemes simple, fasciculate, shorter than the leaves; fruit obovate-obtuse.

Hab.—In mountainous regions. Soumali and Hadramaut.

Mohr Madow. *Playfair.*
Maghrayt d'Sheehaz. *Carter.*

Respecting this tree Dr. Birdwood writes:—" The tree found by Carter growing in " the limestone formation about Merbat and Ras Fartak, the 'Maghrayt d'Sheehaz' " of the Maharas, is probably a smaller tree in habit. The plant which he brought " from the Hadramaut to Bombay is to this day growing there in the Victoria " Gardens: and whilst it is almost impossible to distinguish its leaves from those of " the 'Mohr Madow' of the Soumalis, except from their being smaller and more " crumpled, it is not more than half the height of the plants of Mohr Madow propa- " gated in the Victoria Gardens from the cuttings sent to Bombay by Playfair, and " which are already about 9 feet high. In 'Maghrayt d'Sheehaz' the leaves have " sometimes five pairs only of leaflets."

Playfair's dried specimens of leaves of "Mohr Madow" sent to Kew and Bombay include two very distinct varieties; one crenate, undulate, and pubescent on both sides; the other undulate and obscurely serrulate or almost entire, and velvety and paler below, and glabrous above. The plants of Mohr Madow in the Victoria Gardens, and in Playfair's garden at Aden, have as yet shown only the former variety of leaf, and the plant Maghrayt d'Sheehaz in the Victoria Gardens never has shown any other; but from the analogy of *B. thurifera* and *B. Bhau-Dajiana* (Mohr Add), I am led to believe that the two kinds of leaves included in Playfair's dried specimens of Mohr Madow are only varieties of leaf of one species, and not leaves of two distinct species of *Boswellia*. I have therefore included the description of both leaves in my description of *B. Carteri*, but not without uncertainty.

Boswellia Bhau-dajiana, *Birdw.*

Birdwood in Linnean Transactions, xxvii., p. 144, t. 31.

Small tree; branches downy or smooth at their extremities, leaflets 7–10 pair, oblong-lanceolate or oblong, broadly rounded or truncate at the base, either serrate and pubescent, or obscurely serrulate, and hoary beneath, smooth above; inflorescence in simple fasciculate racemes (immature); fruit oblong, contracted at the base.

Hab.—In Soumali Mountains.

Native name of tree—

Mohr Add. *Playfair.*

To the description of this tree Dr. Birdwood adds the following remarks:—Playfair's dried specimens of leaves of "Mohr Add" sent to Bombay include two distinct varieties, the leaflets in one being serrulate and downy, in the other obscurely and remotely serrulate or entire, and hoary beneath, and glabrous above. The latter is the only variety of leaf sent by him to Kew. But in one of his dried specimens of "Mohr Add" sent to Bombay both these varieties of leaf are seen in the same specimens. The plants in the Victoria Gardens all showed the pubescent and serrate variety of leaf until last year, when one threw out both the varieties of leaf described, and a third variety also of intermediate character. The plant from which I gathered the flower represented in my plate of *B. Bhau-dajiana* showed at that time only the serrate and pubescent variety of leaf; but I have figured beside it two other varieties of leaves, which I subsequently found either on this same plant or other plants of the same species growing near it. It would be more satisfactory if I were less uncertain on this point, especially as my ground for including the two kinds of "Mohr Madow" leaves sent to me by Playfair under the species *B. Carteri* is the all but absolute certainty that analogous varieties of leaf are produced by *B. Bhau-dajiana* or "Mohr Add." It must be distinctly borne in mind therefore that with the leaves no flowers have ever been seen of the variety, either of "Mohr Madow" or "Mohr Add," in which the leaflets are more or less entire and hoary below, and glabrous above, and that it may be an error to include them as varieties respectively of *B. Carteri* and *B. Bhau-dajiana.* But analogous variations of leaf are conclusively authenticated in the case of *B. thurifera.*

The serrate downy variety of leaflet in *B. Bhau-dajiana* is very like the young leaflets of *B. Carteri*, but the leaflets in the latter, young and old, are always more or less undulate, whereas in the former they are always quite flat. In the former, also, the leaflets are narrow and more lanceolate, in the latter broad and more oval; and in the mass of foliage these differences make a distinction.

Soumali Olibanum.

The two preceding species with their varieties appear to yield the Olibanum which is imported into India. Dr. Birdwood believes that his "*B. Carteri* (Mohr Madow) " and *B. Bhau-dajiana* (Mohr Add) yield the bulk of the Olibanum exported from " the Soumali country under the name of Laban Sheheri, possibly (writes Playfair) " because it is principally taken to Sheherr and Makalla markets." Vaughan says Luban Shaharee is the term given to Arabian Frankincense, which is also yielded by *B. Carteri* (Maghrayt d'sheehaz). Both the glabrous-above hoary-beneath variety of *B. Carteri* (Mohr Madow) and of *B. Bhau-dajiana* (Mohr Add) yield according to Playfair the "Luban Bedowi" of the Soumalis, which is also mentioned by Cruttenden, but nowhere by Vaughan.

Laban Sheheri, Luban Shaharee (Vaughan).
Luban Bedowi (Cruttenden), Luban makur (Vaughan).
Luban Marbat or Shaharee Luban (Vaughan).
Luban Berbera, or Mustika (Vaughan)?

The specimen of Luban Makur sent over by Vaughan with his communication is described as "Olibanum in separate opaque, yellowish, rather small tears, to which " bark is frequently attached"; this kind is "from the seaports of Ras Kurree Khor " Bundee, Alholu, Murya, and Bunder Khasoom, in the county of the Worsungali " and Meggertein tribes of Soumalis, who inhabit the extreme north-east coast of " Africa, about Cape Gardafui. The natives collect this gum in the months of May, " June, and July. When picked it realizes 1½ dollars per maund, if not picked about

" half that sum. Very little of this quality of gum finds its way to Aden; almost " all is taken to Maculla and Shehr on the Arabian Coast, from whence it is shipped " to Bombay." (*Vaughan.*)

The " Shaharee Luban " is collected " in the southern and south-eastern districts " of Arabia and exported from several towns on the coast between Ras Tartak and " Marbat. This was the famous thuriferous region, which proved the object of such " diligent search in ancient times. The country still maintains its renown for the abun- " dance of the drug which it yields and for its superior quality, though its value has " sadly depreciated since the time of Pliny, who tells us that those who were employed " in garbling it at Alexandria were hood-winked to prevent their coveting the " precious gum. Three baggalas (Arab trading vessels) are annually freighted from " Marbat to Bombay with an entire cargo of the Arabian frankincense, which realizes " a higher price in the market than any of the qualities exported from Arabia." (*Vaughan.*)

This kind of olibanum, according to the sample which accompanied the communication, consists of tears closely agglomerated together into darkish masses, many of the tears having a vitreous appearance when fractured. Similar, as would appear from the specimens sent, is the " Luban Berbera" or Mustika of Vaughan, which he states " is collected in the district inhabited by the Ayil Yunis and Ayil Hamed " Somali tribes, and upwards of 3,000 maunds are annually sent out of the country. " This quality of gum is generally garbled before it is exported, and is largely used " by the Arabs in their religious services. Its price in Aden is from three quarters " of a dollar to one dollar per maund."

The two species of *Boswellia* named, would appear to be the source of all the Olibanum imported into Bombay. Arabian frankincense probably is wholly derived from *B. Carteri*, it being the " Luban Morbat " or " Luban Shaharee " of Vaughan; whilst African Olibanum, derived from the Soumali coast, may be obtained from both species. At the present it seems impossible to indicate more closely the species or varieties yielding the various varieties of Olibanum. It may be suggested as probable that the Olibanum in tears, the Luban Makur, and the Luban Hunkur of Vaughan, are produced by the same species of *Boswellia*, and the compact or agglomerated kinds represented by Luban Mustika and Luban Shaharee by another species, but of this we have as yet no evidence.

Boswellia thurifera, *Coleb.*

Colebrooke Asiat. Resear. ix. 377, t. 5. *Birdw. in Linn. Trans.* xxvii. p. 146.
Boswellia serrata. *Roxb. Fl. Ind.* ii. 383.
Boswellia glabra. *Roxb. Fl. Ind.* ii. 384. *Roxb. Cor.* ii. 207.

Small tree; leaflets 8–14 pair, dentate serrate or more or less obscurely and remotely serrulate, pubescent or smooth, inflorescence racemose, racemes compact, sometimes dense, much shorter than the leaves; fruit broadly elliptic, triquetrous, base or apex not contracted.

Hab.—Tropical, India.

This is the Indian Olibanum plant which Dr. Birdwood recognizes as a single species; he says, " the natives of India recognise the two varieties of this plant of which Roxburgh made two species, and distinguish between their gum resins. What " I have seen of it has been either stalactiform, like the runnings of a wax candle, " or in small tears, and always so soft that, when kept in a bottle in a short time " it has run into an oleo-resinous mass, with the smell of frankincense but more " turpentiny. I often and perseveringly tried to get regular tears of olibanum from " this plant, but never succeeded in getting anything else than soft oleo-resinous " 'runnings' from it, which, even after months exposure on the trunk, still remained " quite soft. In Khandeish the olibanum produced by this plant is sold under the " name of 'Dup Salai' (*i.e.* incense of Salai) in the village bazaars. But I have never " myself seen it exposed for sale. I have had samples sent to me from all parts " of India, and all soft and always melting at last into a mass." (*Birdwood, Linn. Trans.*)

The only authentic specimen in this Museum is a small sample in the reference collection from the Palamore forests, presented by C. B. Taylor, Esq., in 1845. It is in rather irregular tears of a brown colour, deeper than those of Olibanum in tears in commerce, softer and more friable, easily breaking up into a powder resembling raw sugar. The odour is distinctly that of Olibanum. A sulphur coloured hard resin, more resembling that of *Vateria* than any other Indian resin, was sent over some years since as Indian Olibanum, but it has none of the qualities and properties of Olibanum.

A note occurs in the Journal of Agri. Hort. Soc. of India, vol. i. (1842), p. 290, by Brigadier Twemlow, in which he states that on the Mahabaleshwur hills the Olibanum tree *Boswellia thurifera* grows in great abundance, that it can be propagated from cuttings to any extent, that it produces a gum resin which can be made into tapers, or if made into bricks admixed with charcoal would make a substitute for fuel.

The museum sample of Olibanum from Hatrass is very similar to the one above named from the Palamore forests, and may be derived from the same species.

Boswellia Frereana, *Birdw.*

Birdwood in Linn. Trans. xxvii. p. 146, t. 32.

Small tree. Leaves all smooth and glaucous; leaflets 3–6 pair, broadly ovate, elliptic, obtuse at each end, undulate; inflorescence in compound racemes, much longer than the leaves. Fruit ovate; oblong.

Hab.—Soumali mountain regions.

The gum resin yielded by this tree is called by natives—

Louban Maitee.
Luban Meyti (Cruttenden). } *Ar.*

Of this tree Dr. Birdwood writes,—"This is a strongly marked species, the source of "the 'Louban Maitee' of the Soumalis. It is unmistakeably the plant which Kemp- "thorne describes. It is evidently also Cruttenden's 'Luban Meyti.' Kempthorne "says the tree is one of the most extraordinary plants I ever saw, quite a lusus "naturæ of the vegetable world, for the trees actually grow out of the sides of the "almost polished rocks. The trees were about 40 feet high, the stem was about 2 feet "in circumference, rising straight up, with a bend outwards of 6 or 7 inches. They "are attached most firmly to the rocks by a thick oval mass of substance, about a foot "or so in diameter, something resembling a mixture of lime and mortar. Branches "spring out rather scantily at the top, and extend a few feet down the stem; the leaves "are 5 inches or so long, and 1⅓ broad, narrowing and rounding towards the point, but "not serrated at the edges, the upper surface is of a rich dark shining green, while the "lower is of a lighter hue; they are thin and smooth, and crimped like that beautiful "species of seaweed so often found on the coast of England. The tree has four layers "of bark, the outer being coarse and loose, like that of the beech, while the next two "are as it were glued to the trunk, and delicately fine, resembling oiled paper or gold "beater's skin, and of a bright amber colour; this bark is perfectly transparent, and "can be stripped off easily in large sheets; the natives use it for writing on. The "inner bark of all is an inch or so in thickness, adhering closely to the stem; it is "tough, not unlike leather, and striped red and white, and yields a strong aromatic "perfume. The timber is white, soft, porous, and of little use, except as firewood. A "deep incision into the bark causes the odoriferous gum to exude in large quantities, "which is of a milky white and of the consistence of honey, but it soon hardens by "exposure to the atmosphere. It is a remarkable fact that not a single frankin- "cense tree did I perceive growing upon any other rocks than those of almost "pure limestone. It is my opinion that the trees on the Adel coast are different "from the species as described so ably by Dr. Carter." (*Trans. Bomb. Geo. Soc.* xiii. 1857.)

Vaughan writes:—"The hill ranges on the eastern coast of Africa are composed "entirely of white limestone, in some parts so compact as to resemble alabaster. This "appears to be the soil most genial to the tree, and in no instance did I find it growing "in sand or loam. The tree is first met with at a few miles inland from the coast, "and at an altitude of about 300 feet above the level of the sea. Its appearance is "strikingly singular, seeming at first sight to be destitute of roots, and clinging to the "hard uncreviced rock by masses of a rhomboidal and fantastically shaped wood with "the most obstinate adherence. The stem is nearly at right angles with this substance, "ascending almost invariably in an upright direction, and attaining the height of from

" 12–15 feet. At the base the circumference is equal to that of a man's thigh, " gradually tapering towards the top, where it shoots off its branches and leaves."

" As I saw this plant in Playfair's garden at Aden," adds Dr. Birdwood, " in young " leaf, and covered with bloom, I was much struck by its elegant singularity. The " long racemes of green star-like flowers, tipped with the red anthers of the stamens (like " aigrettes of little stars of emerald set with minute rubies) droop gracefully over the " clusters of glossy glaucous leaves; and every part of the plant (bark, leaves, and " flowers) gives out the most refreshing lemon-like fragrance."

Presuming that the present is the species alluded to by Vaughan, in the remarks above quoted, or that the mode of collection may be similar for all kinds, we may cite in addition Vaughan's description :—" At the proper season incisions are made in the " stem, from which the juice flows forthwith in a copious stream (frequently covering " the entire stem) until the wounds are closed by the dessication of the fluid into a " gum. In this state the trees glisten in their rich investiture, and, as if vexed at " being prevented from pouring forth all their store, the bark distends from the " abundance of sap within. After the juice is inspissated and dried by the action of " the atmosphere and the sun, it is scraped off the trees and the ground beneath, and " collected by the natives, who store it in large loose heaps at particular places on the " seacoast. It is then packed in sheep and goat skins, each parcel weighing from about " 28 to 40 pounds, and transported on camels to the great fair held in Berbera, from " whence it is either sold and then shipped in native vessels for Aden and other ports " on the Arabian coast, or exported to the same places for sale. Sometimes, however, " it is purchased by the Banians, and sent directly to the Bombay market in baggalas." (*Pharm. Journ.* xii. pp. 228.)

The gum resin produced by this tree was sent to England by Vaughan, together with his communication, under the name of Luban Mattee, " so called from Bundur Mattee, " the port from whence it comes. This gum is chiefly collected by the Abardagahala " tribe of Soumalis. The season for piercing the trees from which it is procured is " during the north-east monsoon, in the months of July and August." This sample is now in the Museum of the Pharmaceutical Society of Great Britain, and is just as described by Mr. Dan. Hanbury in a note to Vaughan's communication, where he says, " Luban Mattee is very dissimilar to any resin known in England as Olibanum. It is " in stalactitic masses which have evidently been the produce of a very copious flow of " the peculiar secretion of the tree. These pieces, whose weight varies from one to " three ounces, are in parts white or yellowish, and highly opaque, in other parts " brightly transparent. A thin brown paper-like bark is occasionally adherent. The " Luban Mattee possesses a strong, agreeable, somewhat citron-like odour, and but " little taste. It is closely allied in its characters to the Tacamaque jaune huileuse, " A. of Guibourt (His. des Drogues, iii. 484), which is the Resina anime of the " German pharmacologists. It comes also very close to the Tacamaque jaune huileuse, " B. of Guibourt, a resin of unknown origin, in scraped pieces, which is sometimes sold " in London as Elemi. It also nearly approaches, as Professor Guibourt informs me, " the Resine de Madagascar of his Histoire des Drogues, t. iii. p. 480." This sample, indeed, looks so little like the Olibanum of commerce, and at first sight so much like a kind of Tacamahac, that it cannot be confounded with any other variety of Olibanum. Messrs. Flückiger and Hanbury consider this to have been one of the original forms of Elemi.

A sample of Luban Maitee from Aden was sent to South Kensington in 1873, and is in the Museum Collection.

IMPORTS of OLIBANUM into BOMBAY PRESIDENCY.

—	1869–70.	1870–1.	1871–2.	1872–3.
From	cwt.	cwt.	cwt.	cwt.
Africa, Coast	—	—	5	—
Red Sea	19,298	19,954	22,016	16,837
Aden	778	660	1,213	997
Persian Gulf	1,198	—	1,046	917
Total	21,274	20,614	24,280	18,751
From Madras Presidency	—	—	22	—

EXPORTS of OLIBANUM from BOMBAY PRESIDENCY.

—	1869–70.	1870–1.	1871–2.	1872–3.
To	cwt.	cwt.	cwt.	cwt.
United Kingdom - -	14,360	17,581	14,819	17,446
France - - -	—	—	10	30
Trieste - - -	246	104	—	149
Spain - - -	2	—	—	—
Russia - - -	—	—	335	206
Africa, Coast - -	30	36	12	24
Persian Gulf - -	—	3	49	95
Laccadives and Maldives -	2	6	—	—
Ceylon - - -	—	23	101	202
Straits Settlements -	7	—	164	125
China - - -	2,875	3,841	2,539	6,184
Soumeanee and Meckran -	—	2	—	—
Total - -	17,522	—	18,029	24,461
To				
Bengal Presidency -	—	28	—	41
Madras Presidency -	35	117	219	596
Scinde - - -	4	7	16	2
Total - -	—	—	—	—

Bibliography of Olibanum.

Birdwood in Linnean Transactions, xxvii. p. 111.
Colebrooke in Asiatic Researches.
Wellstead, Travels in Arabia (1838).
Johnston, Travels (1842).
Vaughan in Pharmaceutical Journal, 1st ser. xii. p. 228.
Cruttenden in Trans. Bombay Geogr. Soc. vii. (1846).
Carter in Journ. Bombay Branch Asiat. Soc. vol. ii. 1847.
Flückiger in Schweizer Wochenschrift fur Pharmacie, May 13, 1864.
Royle, Materia Medica, 4th ed. 361.
Pereira's Materia Medica, vol. ii. p. 379.
United States Dispensatory, p. 1569.
Pharmacopœia of India, pp. 52, 443.
O'Shaughnessy's Bengal Dispensatory, p. 283.
Guibourt, Hist. Nat. des Drogues, iii. 473.
Kempthorne in Pharmaceutical Journ. iv. 37.
Flückiger and Hanbury, Pharmacographia, p. 120.

Nagdana.

The Nagdana is a large erect white barked elegant resinous tree (Burseraceœ) with rough compound leaves something in the style of the Cedrelads (writes Mr. Brownlow). The resin, which is a deep transparent red, exudes very freely during the hot months, *i.e.*, March and April (during which months all the resinous trees bleed most freely) and descends into the ground, where large stores are thus formed; long after the tree has disappeared these patches of resin are met with and are dug out, forming the Loban of the bazaar. The ground resin differs from the fresh in being much dirtier, and less transparent; after a fire also, large masses may be collected in the ground. (*Agri. Hort. Journ.* xiii. 351.)

APPENDIX TO GROUP II.

The substances known commercially as Dragon's blood are alluded to here under an appendix, because, although not products of our Indian Empire, they enter to a limited extent into its trade, and therefore cannot be wholly ignored. From their composition they would seem to occupy a position intermediate between Groups II. and III. which this appendix would give them. Their relation to gum resins is somewhat analogous to that of the Kinos to Gums, except that the large proportion of Resin in their composition would favour their retention amongst resinous substances.

Dragon's Blood.

This resin is not the produce of India proper, but is met with as an imported article.

The botanical sources of the different forms of Dragon's Blood are still very obscure. The following plants are named as yielding it :—

Calamus Draco. *Willd.*

Dracœna Draco. *Linn.*

Pterocarpus Draco. *Linn.*

Of these none of the commercial forms are derived from *Dracœna Draco*, L., a native of the Canaries, and it might be excluded from the list of plants yielding the Dragon's Blood of the present day.

Pterocarpus Draco is a tree of the West Indies and South America. There is considerable doubt about its producing any commercial Dragon's Blood.

Lieut. Wellstead has stated that much Dragon's Blood is obtained in the island of Socotra by spontaneous exudation from a large tree, growing at a considerable elevation on the mountains. This remark still requires confirmation, together with more decided information as to the tree as well as its resin.

Hence it seems probable that the Dragon's Blood of modern commerce is produced by some species of *Calamus* and is obtained chiefly from the large islands of the Indian Archipelago.

Ainslie says that this substance "would seem to be often confounded with Kino by " the native doctors, as both on being presented to a Hakeem get the name of *Dum-* " *mulackwayn*, and both on being shown to a Vytian are called *Kanda moorgarittum*. " They mutually consider it astringent, which Dr. Duncan tells us the true Dragon's " Blood is not, and this mistake is, perhaps, owing to the Indian practitioners having, " like Mr. Proust, not sufficiently distinguished betwixt the two substances. I am " inclined to think that genuine Kino is but partially known in the peninsula of India. " The Dragon's Blood which is met with in the bazaars of this country is brought from " Kang Kow, and also from Passier (on the coast of Borneo), where Mr. Elmore in " his valuable 'Dictionary, &c.' (p. 29) says it is to be procured of a finer quality " than in any other part of the world." (*Mat. Med.* vol. i. p. 113.)

Dr. Williams gives Lung-sin-hiang, or "dragon's spittle gum resin" as a name of this drug (in China) placed by him amongst the imports. The tree growing in Sumatra, Java, and other countries to the south of China is said to be met with in the southern provinces of China. Its name is given as Koh-lin, and the lofty tree is described as resembling the *Balsamodendron myrrha*. The tree is said to be chopped to yield this gum, which is supposed to come from Ta-shih-kwoh (Arabia). Dr. Williams describes the drug as in drops of a bright crimson colour when powdered, and semi-transparent. It covers the fruit of the *Calamus draco* and is obtained by beating the fruit in little baskets. The best of this description is imported from Singapore. The common drug of Hankow is sold in large dark red friable masses, which have evidently been packed in matting. It makes a deep blood red, gritty, almost tasteless powder, soluble in spirits of wine. (*Porter Smith, Mat. Med. China.*)

Under the name of Heeraduccan a sample of dragon's blood of superior quality was sent to the Exhibition of 1851 from Sumatra. It was in flat square cakes or bricks of rather more than an inch in thickness, and very hard and compact; when bruised or powdered the colour was more than usually brilliant. It was described as the produce of a large species of rattan growing on the north and north-east coasts of Sumatra, and in some parts of Borneo and imported in small quantities to Bombay.

The remark made by the Rev. F. Mason in his work on Burmah is worthy of note that a species of *Calamus*, which the natives call 'Red ratan,' produces a red exudation like dragon's blood in the forests of Tenasserim. It is a fitting subject for inquiry whether any quantity, and what quality, of dragon's blood is produced in British Burmah or the Malayan Peninsula.

The different varieties of dragon's blood are thus enumerated by Pereira :—

1. *Dragon's blood in the reed;* dragon's blood in sticks. This occurs in dark reddish brown sticks of from twelve to eighteen inches long, and from a quarter to half an inch in diameter, enveloped with the leaf of the Talipat palm (*Corypha umbraculifera*) and bound round with slender slips of cane (probably the stem of *Calamus petræus*). It is supposed to be obtained from a species of *Calamus*, perhaps *Calamus Draco*.

2. *Dragon's blood in oval masses;* dragon's blood in drops. This occurs in reddish brown lumps of the size and shape of an olive, enveloped with the leaf of *Corypha umbraculifera* or *Corypha Licuala*, which thus connects them together in a row, like the beads of a necklace. This kind is rare in English commerce. It is obtained according to Rumphius by rubbing or shaking the fruit of *Calamus Draco* in a bag. A resinous exudation is by this means separated, and is afterwards softened by heat, and made up in these masses.

3. *Dragon's blood in powder.* This is a reddish powder of very fine quality imported from the East Indies. It is probably the dust obtained from the fruit of the *C. Draco* in the way just described.

4. *Dragon's blood in the tear.* It occurs in irregular pieces some as large as the fist. T. W. C. Martius says that pieces of the fruit of the *Calamus rotang* are frequently found intermixed.

5. *Lump dragon's blood.* Dragon's blood in lump varies very much in quality from fine to very ordinary, and occurs in pieces of all sizes and shapes.

6. *Dragon's blood in cakes* occurs in flat, oblong pieces, half the size of a brick, and mostly fine in quality.

Other varieties of Dragon's blood are described, but I have never met with them. Guibourt mentions a dragon's blood in cakes, and a false dragon's blood in oval masses. (*Per. Mat. Med.* ii. 152.)

Crawford says that Dragon's blood is a granular matter adhering to the ripe fruit of a species of rattan (*Calamus draco*) and is obtained by beating or thrashing the fruit in little baskets. Within the Archipelago the principal place of production is Jambi on the north-eastern side of Sumatra. The plant is the wild produce of the forests and not cultivated, although some care is taken to preserve it from destruction. The collectors of dragon's blood are the wild people called Kubu, who dispose of it to the Malays at a price not much exceeding a shilling a pound. The whole quantity produced in Jambi is said to be about 1,000 hundredweights. The article is often adulterated by a mixture of dammer. The best kind imported into Europe in reeds is manipulated by the Chinese. (*Crawford Dicty.* p. 123.)

Dragon's blood is inodorous and tasteless, insoluble in water, but soluble in alcohol, ether, and the volatile and fixed oils, with which it forms red solutions. According to Herberger it consists of—

Red resin - - - - - -	90·7
Fixed oil - - - - - -	2·0
Benzoic acid - - - - -	3·0
Oxalate of lime - - - - -	1·6
Phosphate of lime - - - - -	3·7

(*U.S. Dispensatory*, pp. 1511.)

The museum samples of dragon's blood are a dull red powder from Penang, tears from Bombay and Punjab, lump from Sumatra and Penang, and a fine sample in flattened cylinders 1½ × 1 in. diameter wrapped in palm spathe from Singapore.

Bibliography of Dragon's blood.

United States Dispensatory, pp. 1511.
Pereira, Materia Medica, ii. 151.
Crawford Dictionary, p. 123.
Balfour Cyclopædia, under "Dragon's blood."
Ure's Dictionary of Arts, vol. ii., article "Dragon's blood."
O'Shaughnessy's Bengal Dispensatory.
Smith's Chinese Materia Medica (1871).
Ainslie, Materia Medica, vol. i. p. 114.

Group III.—TRUE RESINS.

Division 1.—COPALLINE RESINS.

In this group are included all those resins which, like copal, dammer, and animé, are readily soluble in turpentine, forming a clear tenacious fluid applicable as a varnish, and which by a more distinct recognition of function might be called "varnish resins." A more close and searching classification would subdivide these for trade purposes into *hard* and *soft*, but a classification of this kind is easily made by the practical man, and is scarcely essential here. There are manifest affinities in the following enumeration, such as the *Hopea* resin with the *Dammara;* the black resin of Travancore with the Pwaingyet of Burma, and the Malayan Rose Dammer; and of the *Shorea robusta* with *S. sericea* and *S. Tumbugaia;* but the more broad and general classification adopted will be found better for practical purposes, than a too minute regard for special affinities.

The resins enumerated here are ranged under the names of their botanical sources, as far as these are known; those of suspected or unknown origin being placed at the end.

Vateria Indica, *L.*

Hook. Fl. Ind., p. 313. *Roxb. Fl. Ind.*, ii. 602. *Corom. Pl.* iii. t. 288. *Wight. & Arn. Prod.* 84. *Wight. Illus.* i. 88. t. 36. *Wall. Cat.*, 3670. *Gærtn. Fruct.*, iii. 29, t. 4.

Vateria malabarica.—*Blume Mus. Bot.* ii. 29. *D'Cand. Prod.* xvi. 2, 624. *Bedd. Ft. Sylv.* t. 84. *Rheede Hort. Mal.* iv. t. 15.

A large tree; young branchlets and inflorescence stellately canescent; leaves 5–8 by 2½–3½ in., oblong or elliptic oblong, obtuse or minutely acuminate, base rounded or emarginate, lateral nerves, 14 pairs, slender; petiole 1½ in.; stipules ½ in., obliquely lanceolate, acute; panicle 6–8 in., terminal, loosely corymbose, branches spreading; flowers ¾ in. diam. 1-ranked, erect; pedicles ½ in.; calyx segments lanceolate, obtuse, canescent on both surfaces; petals elliptic oblong, obtuse, white; anthers glabrous. (Capsule 2½ by 1½ in., oblong, obtuse, coriaceous, fleshy, three-valved.) *Hook. Fl. Ind.* p. 313.)

Hab.—Western Peninsula, from Canara to Travancore.

The resin yielded by this tree bears the following synonyms:—

Sufêd-damar. *H.*
Vellai-kunrikam, Vellai-damar. *Tam.*
Dupa-damaru, Tella-damaru. *Tel.*
Payana, Vella-kunturkkam. *Mal.*
Gum Piney, Piney resin, White Dammar. *Eng.*

The appearances presented by the varieties of this resin are well summarised in the reports of the jury, Madras Exhib. 1857 (pp. 36–55), as follows:—

Var. I.—*Compact Piney resin*, or first sort white dammer.

Occurs in large lumps of all shapes, and varying in colour on the outside from a bright orange to a dull yellow, bearing evident marks of having adhered to the bark of the tree. It has a shining vitreous fracture, is very hard, and bears a great resemblance to amber. Its colour (internally) is of all shades, from a light green to a light yellow, the green tint predominating in the generality of specimens. It is more soluble in alcohol than black dammer, and burns with less smoke and a more agreeable odour.

It is easily distinguishable from all other Indian resins by its superior hardness, its colour, and amber-like appearance.

Var. II.—*Cellular Piney resin*, or second sort white dammer.

Occurs either in small lumps or in large masses generally of a shining appearance and balsamic smell. Has a very cellular structure, which is attributable partly to the mode of collection, and partly to the age of the tree. Notches being cut in the trunk of the tree sloping inwards and downwards, the resin collects in the cavity, and is either permitted to dry on the spot or is collected and dried by the application of heat. It is of all shades from light green to light yellow or white, and is usually translucent. Specimens are sometimes seen in which from the dessication having been improperly conducted the resin is more opaque, of a dull green colour, and full of air bubbles, presenting the appearance of having undergone a partial fermentation.

This resin may be recognised by its cellular appearance and balsamic smell; this latter however (which is of course due to the volatile oil it contains) is gradually lost by long keeping, or constant exposure to the air.

On splitting open old and decayed trees, portions of a dark coloured resin, having the solid consistence of the first variety, but the inferior quality of the second, are often found.

The remark made in the report above cited that "The substance called East India "copal and sold in England as gum *Animi* exudes abundantly from this tree" is a grave error, since the substance known as Animi in England is very different from the Piney resin and is not an Indian product at all, but is derived from Zanzibar. The present substance is almost, if not absolutely, unknown in the London markets.

A communication made to this department gives the following additional details :— "This beautiful substance has long been known, and its properties and local uses "have been repeatedly described. It is also not unknown in England, and I appre- "hend that its cost (and perhaps also ignorance of its peculiar properties) has prevented "its becoming an article of more extended commerce. It should be remarked that "the East Indian Dammer which is well known among varnish makers, though fre- "quently confounded with this, is the product of a very different tree, and is not "produced in this presidency. The finest specimens of Piney resin are obtained by "making incisions in the tree, and are in pale green translucent pieces of considerable "size. The resin that exudes naturally usually contains much impurity. In most of its "properties it resembles copal, but it possesses qualities which give it some advantages "over the latter. Like copal it is but slightly soluble in alcohol, but, as Berzelius "pointed out in the case of copal, it can be brought into solution by the addition of "camphor to the spirit. It is easily soluble in chloroform, and thus might find a "small application as a substitute for amber in photographer's varnish. It differs "most advantageously from copal by being at once soluble in turpentine and drying "oils, without the necessity of the preliminary destructive fusion required by that "resin, a process which tends greatly to impair the colour of the varnish. The solution "of the Piney resin in turpentine is turbid and milky, but by the addition of powdered "charcoal, and subsequently filtering, it yields a solution transparent and colourless "as water and yields a varnish which dries with a purity and whiteness not to be "surpassed. The solution in turpentine readily mixes with the drying oils. It is on "these properties of the resin that its chance of becoming an article of trade will "depend. In price it cannot compete with copal, whose supply to the European "market is regular and abundant. Major Beddome informs me that the cost of "Piney resin delivered on the sea coast would be about six rupees per maund of 25 lbs. "The present price of the best copal in the English market is but 2*l*. 10s. per cwt. Piney "resin yields, on destructive distillation, 82 per cent. of a plurescent oil of agreeable "odour, but not differing essentially from that obtained from cheaper resins."

The additional local uses alluded to for this resin are that "the finest pieces are "called 'kahroba' and sold as amber in the bazaars of Bengal. The resin before it "hardens is the Piney varnish of Malabar; candles are also made of this resin in "Malabar, which Dr. Wight states diffuse in burning an agreeable fragrance, give "a fine clear light with little smoke, and consume the wick without snuffing." It is moreover added that "these candles were at one time introduced into England, but a "very high duty having been imposed the trade ceased."

The fresh resin as it exudes from the tree, called the Pynie varnish, is of a light grey colour with a strong balsamic odour.

Dr. Wight (*Illustrations*) has given the following formula for the manufacture of varnish from Piney resin :—"When used as a varnish the usual practice is to apply "the balsam before it has become hard, but when that is not procurable the resin "melted by a slow heat, and mixed with boiling linseed oils, forms a varnish which "answers equally well for most purposes. As this varnish is of the finest description, "the following formula for its preparation may prove useful to some who are un- "acquainted with the process :—Into a new and perfectly clean earthen vessel put "one part of Piney Dammer in coarse powder, cover closely, and apply a very gentle "heat until the whole is melted; then add about two parts of linseed oil, nearly "boiling hot, and mix well with a wooden spatula. Should the varnish prove too "thick it can at any time be reduced by the addition of more oil, or if required may "be made thicker in the first instance.

"It is essential to the success of the process that the Piney be melted in a covered vessel over a very slow fire, and the whole of it reduced to a fluid before this addition of the oil, which must, to ensure an equal mixture, be nearly boiling hot. The varnish is used for carriages and other fine furniture requiring to have their paint well protected, or to which it is desirable to impart a fine gloss. A spirit varnish is prepared by reducing to powder about six parts of Piney and one of camphor and then adding hot alcohol sufficient to dissolve the mixed powder. Alcohol does not dissolve Piney without the aid of camphor, but once dissolved will retain it in solution.

"The varnish thus prepared is employed for varnishing pictures, &c., but before being used requires to be gently heated to evaporate the camphor, which otherwise will produce by its evaporation a roughness and inequality on the surface of the picture, and spoil its appearance."

There is a peculiarity which is a disadvantage in this resin dissolved in benzole, that it has a tendency to contract and crack in drying. This is manifest if it is employed as a medium, in lieu of Canada Balsam, for mounting microscopic objects, for which purpose the East India Dammer of the London markets is applied. When the latter substance is used advantages accrue which has led to its adoption in preference to Canada Balsam for a large class of objects, whereas, on account of the shrinking in drying above alluded to the resin of *Vateria Indica* cannot be so highly utilized. This is evidence from the nature of the resin itself, if such evidence were required, to refute the assertion still often repeated and affirmed in India that the Piney resin constitutes what is sold in England as East India dammer.

The specimens in the museum are, green from Mangalore and Mysore, yellow from Madras, Coorg, and Travancore.

Vateria acuminata, *Hayne.*

Hayne Arzn, xi. 5. *Hook. Fl. Ind.* 313. *DCand. Prodr.* xvi. 2 (not Thwaites).

VATERIA INDICA.—*Thw. Enum.* 37. *Bedd. Fl. Syl.* t. 14. *Blume Mus. Bot.* ii. 29, t. 4.

ELŒOCARPUS COPALLIFERUS.—*Retz. Obs.* iv. 27.

HEMIPHRACTUM OXYANDRUM.—*Turcz in Bull Mosc.* 1859, i. 262.

A large handsome tree; young branches and inflorescence densely covered with short caducous ferruginous hairs. Leaves 6–8 by 3 in. elliptic, or oblong-lanceolate, shortly acuminate, base rounded, lateral nerves about 20 pairs, stout; petiole 1½ in.; stipules ¾–1 in., linear deltoid, attenuate. Panicle terminal, 8 in., loosely corymbose; bracteoles ⅓ in. ovate, obtuse. Flowers 1 in. diameter, 1-ranked, erect; pedicels ½ in. Calyx ferruginous without, canescent within. Petals orbicular, acute, white. Anthers puberulous. (Fruit 4 in. long by 2½–3 in. broad.) (*Hook. Flor. Ind.* 314.)

Hab.—Ceylon; common in hotter parts of the Island ascending to 2,000 feet.

The resin is known as—

Hal, Hal-dumula. *Cing.*

Thwaites, in alluding to this tree, states that it yields a green resin, used in Cingalese superstitious ceremonies.

The tree seems to be unknown in Continental India, but considerable confusion has occurred with this species and the *Vateria Indica* of Linnæus, the two being perfectly distinct, although the resin afforded by them seems to be almost identical. On the other hand, the Continental *Vateria* seems to be absent from Ceylon, so that there can scarcely ever be any doubt that the Piney resin of Ceylon is the produce of this species, and the Piney Resin of Southern India that of the true *Vateria Indica*.

Vatica lanceœfolia, *Blume.*

Blume Mus. Bot. ii. 31. *Hook. Fl. Ind.* 302. *DCand. Prodr.* xvi. 2, 618.

VATICA CANACA.—*Ham. in Wall. Cat.* 4405. B.

VATERIA LANCEOLARIA.—*Roxb. Hort. Beng.* 42 ?

VATERIA LANCEŒFOLIA.—*Roxb. Fl. Ind.* ii. 601. *Wight. Illus.* i. 88.

VATERIA LANCEOLATA (*Roxb.*).—*W. & Arn. Prod.* 84. *Wall. Cat.* 4405.

RETINODENDRON LANCEŒFOLIUM.—*Korth. Verh. Nat. Ges. Bot.* 56.

SEIDLIA LANCEŒFOLIA.—*Kostel. Allg. Med. Phar. Flora*, 1945.

Young parts pulverulent, but soon glabrous, leaves elliptic, lanceolate, acuminate, base tapering, paler beneath, lateral nerves 15 pairs, ascending.

A middle-sized tree. Leaves 7–11 by 2–2½ in. Panicles two-thirds the length of the leaves. Flowers fragrant. Calyx segments ovate-acute. Petals ⅔ in., cuneate, white. Ovary pubescent; sigma clavate, three-toothed. Capsule ovoid. (*Hook. Fl. Ind.* 302.)

Hab.—Eastern Himalaya, and Eastern Bengal and Sylhet. Khasia Mountains, Assam, and Bhotan.

Moal. *Sylhet.*

Very little is known of the resin derived from this tree. It is stated in Roxburgh's Flora Indica that it exudes a clear liquid from wounds in the bark, which soon hardens into an amber coloured resin from which the natives obtain by distillation a dark coloured, thick, strong smelling balsam, called chooa or chova, by the people who prepare and sell it; and Goond by the Brahmins, who use it in their religious ceremonies and temples.

In Mason's Burmah, however, it is distinctly stated (p. 486) that whilst *Vateria Indica* does not appear to be known in Burmah, another species of the same genus, *Vateria lancefœolia*, yields almost a precisely similar resin. When in bloom the tree is quite ornamental, and diffuses the fragrance of its flowers for a great distance around.

Vatica Roxburghiana, *Wi.*

Wight. Illus. p. 88. *Bedd. Fl. Sylv.* xcv.

Large tree, young branches, petioles, and infloresence minutely furfuraceous; leaves glabrous above, except the costa, ovate to oblong with an obtuse point, rounded or obtuse at the base, 10–14 primary veins on each side; panicles axillary, shorter than the leaves, branchlets few flowered; calyx and outside of petals furfuraceous, divisions of the former acute, enlarging in fruit, then five nerved and glabrescent; petals five times as long as calyx; stamens 15, in two rows; anthers oblong with a short apiculation; style nearly double the length of the anthers; stigma clavate, obscurely 3–6 lobed; fruit globose, not sulcate, crowned with a small nipple-like point at the apex, rough with minute raised reticulations and furfuraceous, hard, woody, and indehiscent.

Hab.—South Canara. Travancore. Ceylon.

Common in some of the Western Coast forests, and produces a gum-resin, according to Beddome, but of its character he gives no further information. The tree is also indigenous in Ceylon, and it is probable that the same species was collected by Griffith in Burmah.

Shorea robusta, *Roxb.*

Gærtn. Fruct. iii. 48, t. 186. *Hook. Fl. Ind.* 306. *Roxb. Fl. Ind.* ii. 615. *Corom. Pl.* iii. 312. *Wall. Cat.* 965. *DCand. Prod.* xvi. 2, 628. *Bedd. Fl. Sylv.* t. 4.
VATICA ROBUSTA. *Steud. Nom.* Ed. 2.

Leaves ovate-oblong, acuminate, base cordate, glabrous; lateral nerves about 12 pairs; panicles terminal or axillary lax; stamens 50.

A very large tree, but often occurring in a stunted form. Leaves 6–10 by 4–6 in. Petiole ¾–1 in. Stipules ⅓ in., falcate, pubescent. Panicles 5–9 in. long, clothed as well as the flowers with pale velvety pubescence, branches unilateral, racemose. Flowers subsessile. Petals about ½ in. long, pale yellow, tapering upwards, 12–13 nerved; anthers, with a bearded appendage. Ovary pubescent; stigma 3-denticulate. Fruiting calyx with the segments, sometimes subequal, bases ⅓ in., ovate, pubescent; wings 2½ in. linear, obtuse, 10-nerved. (*Hook. Fl. Ind.* 306.)

Hab.—Tropical Himalaya and along its base from Assam to the Sutlej. Eastern Districts of Central India. Western Bengal Hills.

Ral, Dhuna, Dhonah, Damar. *H.*
Dhuna, Ral. *B.*
Eing-gyeen. *Burm.*
Guggala. *Can.*
Sal-dammar. *Eng.*
Oanqahar, Qiqahar. *Arab.*
Laale-moaabbari. *P.*
Kungiliyam. *Tam.*
Guggilam. *Tel.*
Rala, Guggilu. *Mahr.*
Guggilam, Koushikaha. *Sans.*

Occurs in brittle stalactitic pieces, of a pale creamy yellow, nearly opaque. Each piece has a striated appearance as if the resin had run out in thin liquid streams, which had coagulated on the surface one over the other. (*Punjab Products*, p. 411.)

In O'Shaughnessy's "Dispensary" it is observed that Capt. Jenkins, of Assam, sent specimens of this resin, which were perfectly transparent and colourless, characters quite at variance with the ordinary appearance of this resin.

The specific gravity of the resin is stated to be 1·097 to 1·123, easily fusible, partially soluble in alcohol, almost entirely so in ether, perfectly in oil of turpentine and the fixed oils; sulphuric acid dissolves and gives it a red colour. "Two parts of colourless "dammer and 2½ parts of oil of turpentine make the best varnish for lithographic "drawings."

Sal Dammer is dissolved much more freely and speedily in benzole than in spirits of turpentine. The solution in both cases is turbid and unpromising, so that there is but little probability of its commending itself as a varnish resin. The solution in turpentine makes moderately good tracing paper, and it may be employed as a paper varnish, but it is doubtful whether it possesses any peculiar advantages for such a purpose. As a paper varnish it is clear, nearly colourless, and dries rapidly without cracking.

Specimens in the Museum are from various localities, as Gyah, Raepore, Cuttack, Beerbhoom, Chota-Nagpore, Lucknow, Vizagapatam, Tenasserim, Johore, Malacca, and three from Burmah. Although externally closely resembling each other it is possible that some of the latter from beyond the Ganges may have been collected from allied species.

Shorea sericea, *Dyer.*

Dyer in Hook. Fl. Ind. 306.

A very large dammar-producing tree; young parts, nerves of the leaves beneath, and inflorescence covered with rusty meal intermixed with fascicles of short hairs; leaves $4\frac{1}{2}$–7 by $1\frac{3}{4}$–$2\frac{3}{4}$ in. long, polished above, lateral nerves very prominent beneath, petiole $\frac{1}{2}$–$\frac{3}{4}$ in.; panicles about 6 in. long, ultimate branches racemose, uni-lateral; flowers sessile, buds clothed with tawny silky hairs; anthers with the lobes sub-equal; stigma 3–lobed; fruiting calyx with immature accrescent segments, clothed on both surfaces with soft yellow down. *Hook. Fl. Ind.* 306.

Hab.—Malacca.

Dammar-batu. *Malay.*

Of the produce of this tree nothing seems to be known in this country, except the fact that the tree yields a kind of dammar supposed to be one of the varieties of Dammar-batu, of which some closely resemble the resin of *Shorea robusta.*

Shorea Tumbuggaia, *Roxb.*

Roxb. Hort. Beng. 42. *Fl. Ind.* ii. 617. *Hook. Fl. Ind.* 306. *DCand. Prod.* xvi. 2, 630. *Beddome Fl. Sylv. Gen.* 26.

SHOREA PENICILLATA, *DCand. Prod.* 630.

VATICA TUMBUGGAIA, *Wi. & Arn. Prod.* 84. *Wight Ic.* t. 27.

Leaves ovate or oblong-cordiform, acuminate, base truncate or emarginate, glabrous on both surfaces; lateral nerves about 8 pairs; panicles terminal; stamens about 30.

A gigantic dammar-producing tree; leaves $2\frac{1}{2}$–8 by $1\frac{1}{2}$–$4\frac{1}{2}$ in.; petiole 1–2 in.; panicles 8 in. long, hoary or nearly glabrous; flowers shortly pedicelled, buds densely hoary; anthers with a hairy appendage; stigma 3-lobulate; capsule $\frac{3}{4}$ in. long, ovoid, acuminate, pubescent above; bases of fruiting calyx segments $\frac{1}{2}$ in. long, ovate, hoary; wings $1\frac{1}{2}$–$1\frac{3}{4}$ by $\frac{1}{2}$ in., spathulate, obtuse, 8–10 nerved. *Hook. Fl. Ind.* 306.

Hab.—Western Peninsula. Forests of Cuddapah and Palghaut in Mysore.

Tambugai. *Tam.*

The India Pharmacopœia states that a portion of the black dammar of the Bazaars is said to be the produce of this tree, which is probably an error, as the resin of Tambuggaia is scarcely distinguishable from "Ral."

A correspondent quoted in the Madras Jury Reports for 1857 states that there are extensive tracts of Googulam jungles in Goomsur and Cuttack provinces. The Khoonds and Woodias living in and near these jungles wound trees in several places. The resin issues, and is collected when sufficiently solid. The dammer collected from the decayed parts of the tree is of a dark colour; the tree is called "Guggilam" in Telugu, and "tala gotso" in Uriya. The Khoonds and Uriyas make the leaves into plates from which they eat their food, and also roll up tobacco in them to smoke like a cheroot. In time of famine the above tribes live on a soup made from the fruit of this tree. The *Shorea Tumbuggaia* grows also to a limited extent on the West Coast, but yields little, if any, of the dammer collected there.

Hopea odorata, *Roxb.*

Roxb. Hort. Beng. 42. *Flor. Ind.* ii. 609. *Hook. Fl. Ind.* 308. *DCand. Prod.* xvi. 2, 632.

HOPEA FAGINEA. *Hort. Calc.*

HOPEA EGLANDULOSA. *Roxb. Hort. Beng.* 42. *Flor. Ind.* ii. 611. *DCand. Prod.* xvi. 2, 634.

HOPEA DECANDRA. *Buch. ex Wight Ill.* 88.

HOPEA WIGHTIANA. *Miq. in Herb Hohen,* 320.

Leaves ovate or lanceolate-oblong or obtusely acuminate; lateral nerves about 12 pairs, rather weak, their axils often glandular; anthers oblong; appendage about the same length.

A large tree; branchlets dark brown, glabrous; leaves 4–6 by $1\frac{1}{2}$–2 in.; base rounded, often sub-oblique; margins undulate, firm, glabrous on both surfaces; petiole $\frac{3}{4}$ in.; flowers sweet-scented, buds about $\frac{1}{8}$ in., very shortly pedicelled; calyx-segments ovate, obtuse; petals yellow; nut $\frac{1}{4}$ in. ovoid, apiculate, glabrous; wings of fruiting calyx-segments linear, obtuse, tapering to an oblong base $\frac{1}{5}$ in. (*Hook. Fl. Ind.* 308.)

Hab.—Rangoon. Pegu. Martaban and Tenasserim.

Thengan-tsee. *Burm.*

The resin derived from this source occurs in nodules about as large as a walnut, rounded, of a pale straw colour, sometimes almost colourless, brittle, with a shining resinoid fracture, scarcely distinguishable in appearance from the East India dammer of the London markets.

In the Journal of the Agri Horticultural Society this resin is stated not to be soluble in spirits of wine, but perfectly so in benzole and turpentine. The varnish afforded by both is clear and limpid, and dries almost instantaneously. (*Proceed.*, vol. x. p. cxxvii.)

We have found this resin to resemble greatly the East India dammar (so called in the London market) from Singapore, which is referred to *Dammara orientalis.* It dissolves with equal freedom in spirits of turpentine or benzole, producing a clear bright solution, which when used as a varnish dries rapidly and smoothly. In all essential qualities it seems to equal East Indian dammar, even for microscopical purposes, and hence gives good promise of taking a similar place and rate. Externally it has also sufficient resemblance to make it difficult to discriminate the one from the other. The advantage of hardness is rather in favour of the Thengan. The London brokers class it as a copal, and value clear pale samples at a rate equal to that of East Indian Dammar, about 40*s.* per cwt.

In the India Pharmacopœia it is stated that this resin reduced to powder forms a popular styptic amongst the Burmese, its action probably being purely mechanical.

The samples in the Museum Collection are from Burmah and Pegu, and one sample from the former locality is mixed with fragments of a much darker colour, and altogether much inferior in quality.

Hopea micrantha, *Hook.*

Hook. f. in Trans. Linn. Soc. xxiii. 160. *Hook. Fl. Ind.* 310. *DCand. Prod.* xvi. 2, 634.

Branchlets dark coloured, minutely lepidote; leaves elliptic-ovate or lanceolate, candate-acuminate, obtuse; lateral nerves very numerous, evanescent; panicles terminal or axillary, less than half the length of the leaves, hoary-pubescent; calyx-segments varnished.

A very lofty tree; leaves 2–$2\frac{1}{2}$ by $\frac{3}{4}$–$1\frac{1}{4}$ in., glabrous on both surfaces, mid-rib prominent above and beneath; petiole $\frac{1}{2}$ in., puberulent; flowers $\frac{1}{10}$ in.; pedicels fully $\frac{1}{12}$ in.; petals palely pubescent; anthers sub-orbicular, appendage about as long; nut conical; bases of fruiting calyx-segments $\frac{1}{4}$ long, with two gibbosities at the bottom; wings $1\frac{1}{2}$ in., oblanceolate, obtuse, 9 nerved. (*Hook. Fl. Ind.* 310.)

Hab.—Malacca. Borneo. Labuan.

Dammar-mata-kooching, Dammer-batu. *Malay.*

Dr. Hooker, when describing this tree, says:—"Motley's specimens vary a little in "the breadth of the leaf. He sends them under two names, viz., as the rock "dammer, a very large tree with furrowed bark, hard wood turning brown when cut, "yielding sparingly a white resin and one of the best timbers; also as the tree yield-"ing Dammar-mata-kuching of no great diameter, but tall with smooth buttresses, "deeply fissured bark on the trunk, wood hard, light yellowish brown, durable, easily "worked; the resin exudes in large lumps from the trunks and branches."

The resin occurs in rounded or irregular pieces seldom larger than a walnut, externally amber coloured, but darker and not so friable as the similar resin from *Hopea odorata.* Shining and vitreous in fracture, and internally of a bright straw colour, sometimes lemon-yellow; soluble in turpentine or benzole, forming a clear limpid varnish.

The samples in the Museum which are referred to this species have been received from Assam (Seebsaugor), Johore, Singapore, and Sumatra.

Hopea parviflora, *Bedd.*

Bedd. Fl. Sylv. t. 7, *Gen.* xxvii. in part. *Ind. For. Rept.* 1863–4 with plate. *Hook Fl. Ind.* 308.

Leaves ovate or oblong, acute, lateral nerves about 10 pairs, often forked, their axils often glandular, tips of petals dilated and crisped, anthers sub-orbicular, appendage about half as long again.

A large tree; branchlets reddish brown, slightly glaucescent. Leaves about $3\frac{1}{2}$ by $1\frac{1}{2}$ in., firm, base acute or rounded, margins slightly undulate, glabrous on both surfaces, nerves pale beneath, petiole $\frac{1}{3}$ in. Flower-buds about $\frac{1}{8}$ in. tomentose, very shortly pedicelled; calyx-segments lanceolate, obtuse. Fruit unknown. (*Hook. Fl. Ind.* 1309.)

Hab.—Malabar and S. Canara, ascending to 3,500 feet.

Kiral-boghi. *Can.*

The only specimen of this resin in the collection is a black pitchy looking resin in irregular pieces with a dull opaque fracture, and not at all resembling the produce of the two species of *Hopea* already named. This specimen was received from Madras.

Canarium strictum.

Roxb. Fl. Ind. iii. 138. *Drury Hdbk.* i. 248. *Wi. & Arn. Prod.* 175. *Dalz. Bomb. Fl.* 52.

Tall straight tree; young parts densely clothed with a rusty coloured pubescence; leaves pinnated, very long; leaflets 4 pairs and an odd one, ovate-oblong acuminated, slightly crenulated, hairy beneath on the nerves, 5–12 in. in length.

Hab.—Tinnevelly. Meera Hills.

The native names are—

Kala-damar. *H.*
Karuppu-damar. *Tam.*
Nalla-rojan. *Tel.*
Kala-damar. *B.*
Kalo-damar. *Guz.*

The strength of evidence is in favour of this being regarded as the source of at least the greater portion of the black dammar of India. Wight observes that the *Canarium strictum* of Roxburgh is known in Malabar under the name of the black dammer tree in contradistinction to the *Vateria* or white dammer tree. This tree is rather common in the Alpine forests about Courtalum in the Tinnevelly district, and is there regularly rented for the sake of its Dammer. This dammer is transparent and of a deep brownish yellow or amber colour when held between the eye and the light, but when adhering to the tree has a bright shining black appearance. (*Illustrations.*)

In the Jury Reports (Madras Exhibition, 1857) the opinion of Dr. Waring is cited in favour of this determination. He says "I regret that there are no flowers to the " specimens of the black dammer tree, but the fruit and the arrangement of the " leaves at once prove that it is not a Dipterocarpus, and consequently not the *Vatica* " *tumbugaia*, whilst at the same time it appears to me clearly to be a *Canarium*." A subsequent note affirms that the specimens had been identified with the *Canarium strictum*, Roxb.

Mr. I. Brown of Trevandrum, in forwarding the specimens to the Madras Exhibition accompanied them by the following remarks:—"It seemed to me that if there was " still a doubt as to the black dammer tree it must have been due to want of authority " in the specimens produced, and thinking that I could not help you well by such " proxy as I could have obtained, and as I expected to be down here near the base " of the ghauts and in the region of dammer trees within a fortnight, I the more " readily postponed the matter. The black dammer tree was in flower about the " beginning of April as far as I could guess from the seed vessels, and the white " dammer tree in the beginning of May. Both trees grow in the forests here about " 1,800 feet above the level of the sea, but the white dammer tree seems more " common than the other, perhaps because the hill men getting more dammer from " the latter destroy it more readily. The best specimens of the black dammer tree " which I examined were about two yards in girth at the height of four feet above " the spread of the roots. The trunk is round, straight, and smooth, rising 20 to 30 " feet before branching, the bark generally whitish dotted with small papillæ, peels " off in long flakes. The dammer that exudes from the cuts in the trunk seems to be " a great favourite of several species of insects, especially of one resembling a bee, "called by the hill men "kulliada," (earth hole) which live in pairs in holes in the "ground. I send you specimens of the bark as well as of the black dammer taken "by me from one of the trees. The hill men to get the dammer make a great "number of vertical cuts into the bark all round near the base of the trunk, they "then set fire to the tree below the cuts, and having thus killed it, they leave it for "two years before they collect the dammer, they say that after one year only the "quantity of dammer is much less than after two years. The tree is killed in the "hot season and the dammer is collected in February or March. When on the "ghats previously, as well as this year, we were struck on looking towards the forests "on both the Eastern and Western slopes as high as 3,000 feet above the level of the "sea, by numbers of trees with bright red, often crimson, foliage, contrasting strongly "with various greens around. These crimson trees are black dammer trees; the "colour due to the young leaves disappears gradually in April. The leaves of the "young plants are more sharply serrated than those of the old trees, and they have "tufts of hairs all around the edges, which are not visible in the leaves taken from "the old trees."

The assistant supervisor of the Coimbatore district also confirms the method employed for obtaining the resin. "The dammer is extracted from the black "dammer tree by piling firewood to the height of one yard round the base of the "trunk and lighting it. The dammer subsequently exudes from the trunk as high "as the flames reached. The operation is conducted at any season of the year, and "the dammer continues to flow for ten years between the months of April and "November and is collected in January. The tree after having yielded dammer for "10 or 12 years decays."

The quantity obtained from a black dammer tree is stated to be about 30 to 40 dungallies, the dungally being equal to about 2½ seers.

Black dammer occurs in large stalactitic-shaped masses of a bright shining black colour when viewed from a distance, but translucent and of a deep reddish brown when held in thin laminæ between the eye and the light. It is perfectly homogeneous, and has a vitreous fracture. Its shape appears to be due to the fact of the balsam having exuded in a very fluid state and trickled down the trunk of the tree, where it is gradually hardened by exposure to the sun; the fresh resin continuing to flow over that already hardened gives rise to the stalactitic appearance of the huge lumps of resin, the outside of which much resemble the guttering of wax caused by placing a lighted candle in a draught. It is insoluble in cold, but partially soluble in boiling alcohol on the addition of camphor; when powdered it is readily soluble in oil of turpentine. Powdered and burnt on the fire it emits a more resinous smell, and burns with more smoke than white dammer. The size of the lumps of this resin, together with its colour and the peculiarity of shape already mentioned, suffice to distinguish it from other Indian resins. (*Reports, Madr. Ex.* 1857.)

In a communication to this department the following remarks are made on this resin. "This well known substance offers little chance of usefulness, in Europe at "least, when the many resins are considered that are found in the market at a far "less price. It is used in this country for many small purposes, as in the manufacture "of bottling wax, varnishes, &c. Its colour when in solution is pale, if compared "with its dark tint when in mass. Thus though insoluble in spirit its solution in "turpentine forms a tolerable varnish. When submitted to destructive distillation "it yields about 78 per cent. of oil resembling that obtained from common colophany, "but I fear in the majority of its possible applications it possesses few advantages "over ordinary resin at 7*s.* 6*d.* per cwt. Major Beddome estimates the price of "black dammer on the coast of Canara at 8 Rs. per 25 lbs. (or nearly ten times the "price of resin in England). The number of substances suitable for coarse varnishes "have lately become very numerous in Europe. Common resin is now purified by "a patent process consisting of distillation with superheated steam, by which it is "obtained nearly transparent and colourless as glass in such amount that a single "firm turns out 60 tons per week."

The uses of this resin in medicine are limited, and mostly as a substitute for Burgundy pitch. Dr. Bidie (in Madras Quart. Jour. Med. Sci., Oct. 1862, p. 292) gives formulæ for Emplastrum Picis; Emplastrum calefaciens; and Emplastrum Opii, into which this dammer enters as an ingredient.

The samples of this resin in the Museum collection are from Madras, Cochin, two from Coorg, Cudoor, Seebsaugor, Tenasserim, Malacca, and Burmah. It is very doubtful whether all of these, especially the latter, are derived from the same species.

Poon-yet. Pwai-nget.

It is generally supposed that at least a portion of the peculiar honeycombed black resin of Burmah which is known by the above names is derived from the black dammer tree, *Canarium strictum*. Roxb.

In Mr. Brown's account of the black dammer already cited he states that "it "seems to be a great favourite of several species of insects, especially of one "resembling a bee, called by the hill men "kulliada," which live in pairs in holes "in the ground." This remark is important in connexion with what follows here on "Poon-yet."

In 1855 a substance from Malacca was exhibited at the Madras Exhibition under the name of dammer klotee, described as having a peculiar honeycombed structure, although it is remarked in a note to the Jury Reports, 1857, that the cells were larger and the resin blacker and harder than Poon-yet.

The samples of Poon-yet from Burmah exhibited at Madras were said to be found in the ground, and were supposed to be produced by some hymenopterous insect, it is also found in hollow trees. This substance, although composed chiefly of a yellow brittle resin, mixed with more or less sand or earth, has an appearance precisely like that which would be looked for if used by an hymenopterous insect in the construction of its cells. (*Madras Reports, Exh.* 1857.)

From the above it would be concluded that Poon-yet is a black resin, resembling black dammer, with a peculiar honeycombed structure, which structure is more or less due to the manipulation of a species of dammer bee, or hymenopterous insect, particulars with which specimens of Poon-yet from Burmah coincide.

Under the head of dammer bee in Mason's Burmah (2nd Ed., p. 350), it is observed that "there are several species of bees that are characterised by building their nests "in the cavities of trees. The wax of one species is dark coloured, of the consistence "of resin, and is much used by the Burmese in caulking boats. It is constantly seen "for sale in the bazaar, and Europeans generally regard it as the resinous product of "a tree, or dammer." So Dr. Helfer considered it, and recorded it "a kind of balsamic "gum resin." Three years after the publication of the above, Dr. McClelland wrote in his report on the teak forests of Pegu "Dammer, *Pwai-ngyet*. This article is found "in the bazaars throughout Pegu. It is yielded by the *Shorea robusta* (manifestly "an error), or the saul tree of India, which occurs plentifully in the forests on the "Shan side of the Sitang, east of Toungoo, and also, but to a less extent, in the "forest of the Prome District. I have not found, however, that the dammer is "obtained from these forests, as none of the trees appear to have been perforated for "it, so the supplies of this article are probably brought down to Pegu from forests "beyond the frontier." Three years more elapse, and the Rev. C. S. Parish writes me, "the secretary of the Agri and Horticultural Society of Calcutta wrote to me "some time ago for information about the pwai-ngyet, which several people con- "founded with dammer, and take to be a gum. I sent up some specimens of "pwai-ngyet, and also several of the insects, which were forwarded to England, to "Mr. Fred Smith, of the British Museum, who has identified it as *Trigona læviceps*, "first sent from Singapore."

In reference to this substance, a letter was addressed by the Rev. C. S. Parish to the editor of Science Gossip in 1866 (p. 198), in which he writes :—" A few years ago the secretary of the Calcutta Agri-Horticultural Society wrote to me for information concerning Pwai-ngyet. Until then I knew very little about it, but my attention having been called in this way to it, I made it my business to find out what it was. If I read your remarks rightly, you appear to be of opinion that Pwai-ngyet is the pure unadulterated gum or resin of *Canarium strictum*, only bored and channelled by the bee. If this were so, then the substance should only be found on that tree. It is however found on different trees; sometimes too in the ground, or in a hollow among rocks, and occasionally even in the hollow post of an old house. I have seen the bees making their nest in all these several situations. Pwai-ngyet, I believe myself, is a combination of various gums or resins, and probably also of oils, gathered from various sources while in a soft state by the bee, and built up and moulded very much as wax is moulded, with this difference, that whereas wax is formed by the honey bee into cells of perfect and uniform symmetry, the cells in Pwai-ngyet assume no regular form at all. What trees contribute their juices to form Pwai-ngyet I cannot say for certain, though I incline to think that Thengan tree, or the resin of *Hopea odorata* is the chief ingredient, and that the oil of the various Dipterocarps or wood oil trees, particularly of *Dipterocarpus lævis*, the wood oil tree par excellence, also enters into the composition of the material. My reasons for thinking so are, that the texture, the colour, and the smell of *Pwai-ngyet* are all such as would apparently result from a combination of the two substances mentioned, and that *Hopea odorata* and *Diptero-carpus lævis* are among the principal giants of our forests and common trees. On the other hand, I do not think (though I will not be sure) that *Canarium strictum* is found in our provinces, although I believe a species of *Canarium*, a large timber tree, is found in Pegu.

"The *Trigona læviceps* builds its nest generally in the hollow of a tree, entering by a small aperture. These apertures are lined with *Pwai-ngyet*, and sometimes only show a small rim of that substance raised above the bark of the tree. Sometimes, however, (perhaps always if undisturbed), the bees go on building outside and adding on to the rim until they have formed a wide mouthed entrance which projects as much as a foot from the tree. These structures commonly assume the shape of the mouth of a large trumpet flattened horizontally, and have a perpendicular diameter of a foot or so, and a horizontal diameter of three or four inches. They are built with great regularity in their exterior half, but not so regularly towards the base, from the necessity of adapting the structure to the shape of the tree where the hole may chance to be. They are very curious and pretty objects, but being very prominent, attract the notice of the passer-by, and so often lead to the spoiling of the habitation. I send you a rude sketch of one of these trumpet openings. I despatch also a small piece of Pwai-ngyet broken off the upper base of such a work. By holding this up to the light, you will see three or four large cells of about an inch in diameter without any opening. I can only suppose that the object of these cell walls is to strengthen the narrow base in its support of the larger projecting mass. If so, here is another instance of a mysterious intelligence possessed by one of the smallest of living creatures. The piece marked No. 1 will show the ordinary texture, colour, and general appearance of Pwai-ngyet as it is found in the jungles. No. 2 is nearly white, a very unusual colour. I send it because of its greater resemblance to *Thengan-tsee*, or the resin of *Hopea odorata*, of which I send also a small piece in order that you may be able to ascertain how far the two substances are chemically identical. What the internal economy of the nest of *Trigona læviceps* is I cannot say, as the tree has commonly to be felled in order to obtain the contents, and this I have never seen done. I am informed by the Burmese that from five to ten viss are usually obtained from one nest. A viss is about 3⅗ lbs., and costs about 4 annas (sixpence) in the bazaar. I should imagine that, considering the source whence it is procured, the supply must be very limited, and if exported it would soon equal bees' wax in price. Its principal, if not only use, at present is for caulking; and for this purpose it is mixed with earth, oil, or petroleum. The method is to boil the Pwai-ngyet in water, which makes it quite soft, and then to knead it with a certain quantity of petroleum, until it attains the consistency of a lump of putty, which it much resembles. In that state it is fit for use, and is extremely viscid and tenacious. On putting a piece of Pwai-ngyet into boiling water in order to perform the operation myself, I noticed that the surface of the water was covered with a thin film of oil. This confirms me in the idea that oil is united with resin in the composition of Pwai-ngyet. It is soluble in oils and in turpentine, but not in spirits of wine. I may conclude by mentioning that the mistakes made by several persons in Burmah with regard to the origin of this substance must have arisen from the fact that the name of *Pwai-ngyet* is often used here in the bazaars, to denote any kind of resin or dammer, but the true *Pyai-ngyet* of the Burmese is that made by the small bee called *Trigona læviceps*, and is made by them in the manner I have tried to describe.

Fig.

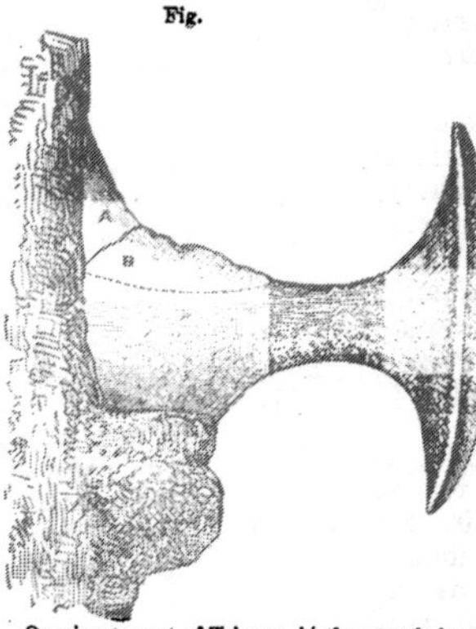

Opening to nest of Trigona, 1/6th natural size.

Fig.

Front view of mouth.

"The drawing, one sixth the natural size, is an imaginary restoration, as but the lower half is before me. The portion marked A forms specimen No. 1 (above alluded to); B within the dotted line marks the space over which the blind cells, for strength as I suppose, extend. From the shape of the base as brought to me the mass must have rested partly on some such excrescence of the trunk as I have tried to represent The actual entrance into the tree in this instance was by a narrow perpendicular slit two and a half inches long and three tenths of an inch wide, the upper part of which may be seen in the specimen. The width, laterally, in the middle of the stem of the structure, is exactly one and a half inch. The weight of the whole, judging by the portion I have, may have been half a pound."

From this letter, and in default of further information of the bulk of material within the body of the tree, it may reasonably be assumed that, in Burmah, the name of *Pwai-ngyet* is applied to three substances, at least,—

I. True *Pwai-ngyet*, which is a structure formed by the dammer bee for its own habitation, probably from a mixture of substances, and is found on various trees and in numerous situations, as in the ground, rocks, &c.

II. Honeycombed or perforated black resin, with a strong elemi odour, found in amorphous masses, often very large, and which *may* be the internal structure above alluded to, although this seems doubtful.

III. Black dammar with a vitreous fracture and without odour, or any perceptible modification of its structure which can be referred to external interference, such as orifices or channels.

If the observation be confirmed which is made by the Rev. C. Parish, of Moulmein, in the above-cited communication, then it is certain that none of the black dammers of Burmah are the produce of the *Canarium strictum*, and it is still a desideratum to ascertain what trees produce the black dammar commonly found in the bazaars.

The samples in the Museum collection from Burmah differ from the ordinary black dammer of Southern India, in being less brittle, and making the fingers sticky after handling them, in having a more earthy and less shining fracture, and in a more decided and often very distinct and strong balsamic, elemi-like odour when rubbed or broken. For these reasons also it is open to doubt whether the Pwai-ngyet is really derived from *Canarium strictum*.

Samples of this honeycombed resin from Burmah are found to be partly soluble in spirits of wine and partly in turpentine, but not wholly in either. It appears to consist of a dark coloured resin which is soluble in cold spirits of wine, and a pale resin which is disintegrated but not dissolved by spirits of wine but soluble in turpentine. In this respect this resin differs from the black dammar of Travancore, and a similar black dammar from Burmah, which are not at all soluble in cold alcohol, but dissolve at once and entirely in turpentine. This kind of Pwai-ngyet is very similar to the Canara resin (*Bursera acuminata*) and the Peraman resin of Venezuela partaking more of the character of elemi than of copal.

Canarium Bengalense, *Roxb.*

Roxb. Fl. Ind. iii. 136. *Drury Hdbk.* i. 248.

Large tree; leaves alternate, unequally pinnate; leaflets subopposite, 6–10 pair, from ovate-oblong to lanceolate, acuminated, entire; stipules subulate, clothed with long, soft, ferruginous down; racemes solitary, supra-axillary, somewhat panicled; drupe oblong, size of a large olive, smooth, filled with firm yellowish pulp; nut obscurely three-cornered, almost covered with a thick and hard aril, 3-celled; seeds, one in each cell; flowers small.

Hab.—Sylhet.

This species yields a large quantity of very pure clear amber coloured resin, which soon becomes hard and brittle, and is not unlike copal, yet the natives set little or no value upon it. In the Calcutta bazaar it sells at two to three rupees per maund of 80 lbs. It is a native of Sylhet and the adjacent mountainous countries, and flowers in May and June. (*Madras Jury Reports*, 1857.)

No specimen of this resin has been received for any of the European exhibitions, and none is to be found in the collections of the Museum.

Dammara orientalis, *Lamb.*

Lambert Pinet. ii. 70, t. 43. *Carrière Conif.* 618. *Loud. Arbor.* iv. 247, f. 2308–9. *Loud. Encycl.* 1066. *Forb. Pinet. Wob.* 169, t. 58. *Endl. Syn. Conif.* 189. *Lindl. & Gord. Journ. Hort. Soc.* v. 221. *Knight Syn. Con.* 45. *Gord. Pinet.* 79.

Dammara alba. *Rumph. Herb. Amb.* ii. 174, t. 57. *Blum. Rumph.* iii. 212.

Pinus dammara. *Lamb. Pinet.* i. 61. t. 38. *Voigt. Syl.* ii. 53.

PINUS SUMATRANA. *Hort. Belo. Mirb. Mem. Mus.* xiii. 69.
ABIES DAMMARA. *Poir. Dict.* v. 35.
AGATHIS LORANTHIFOLIA. *Salisb. Linn. Trans.* viii. 312, t. 15. *Blume Enum. Pl. Jav.* 90.
AGATHIS DAMMARA. *Rich. Conif.* 83, t. 19.
DAMMARA LORANTHIFOLIA. *Spach. Hist. Veg.* xi. 336.

A huge tree; leaves opposite, but sometimes alternate, ovate-oblong, attenuated at the base, obtuse or rounded on the point, quite entire, glabrous, of a thick leathery texture, and glaucous green, from 2–4 in. long, nearly $1\frac{1}{2}$ broad at the widest part, straight, rarely falcate, smooth, and dull green on both faces, somewhat two-rowed on the young branchlets and distant, those on the young plants almost lanceolate and sharp pointed; branches vertical, a little reflected, and ascending at the extremities, forming a small head on adult trees; branchlets and lateral branches in opposite pairs and spreading; male catkins about 2 in. long; cones globular or turbinate singly, on footstalks, rising from the axils of the leaves near the extremities of the branches from 3–4 in. long and more than 2 in. broad; scales adpressed, smooth, rounded at the top, thick, and very closely inlaying; seeds unattached with an obtuse one-sided wing, covering the rib of the scale.

Hab.—Moluccas. Sumatra. Java. Borneo. Amboyna. Ternate.

Dammar, Dammar puti, Dammar qatu. *Mal.*

"Le dammara orientalis produit en grande quantité une résine d'abord transparente, molle et visqueuse, qui répand, à l'état liquide, une odeur aromatique qu'elle perd en partie par la dessication. Pour arriver à ce dernier état elle passe graduellement du blanc primitif à la couleur du succin." (*Carr. Conif.* 619.)

To this brief account we may add from Gordon that, "the timber is of little value, "but it produces a fine transparent resin, sometimes hanging like icicles, and much "esteemed by the natives for incense." This is undoubtedly the substance which comes to England by way of Singapore and is sold under the name of East India dammer, being well known as such amongst varnish makers, and the name is not applied by them to any other substance.

It occurs in nodules, more or less rounded, of very variable size, some samples scarcely exceeding that of an egg or a walnut, whilst in others attaining a diameter of several inches. In colour externally whitish and pulverulent with the minute dust of mutual attrition, beneath which it is straw coloured or very pale amber, rather brittle and breaking with a shining vitreous fracture, very clear and transparent, and almost, if not quite, inodorous. Easily and entirely soluble in benzole, ether, or chloroform, less rapidly so in turpentine, forming a clear, limpid, nearly colourless varnish which dries rapidly on exposure to the air.

The Museum specimens are from Singapore and the Malay Peninsula.

This is probably the "soft Indian copal" of Filhol (in *Pharm. Journ.* ii. 773), but not the "oriental copal" of Landerer (*Pharm. Journ.* v. 186).

Trachylobium Mozambicense, *Peters.*

Peters Mossamb. t. 2.

Hab.—Eastern Africa

Bombay Animé, E. Indian Animé, Gum Animé. *Eng.*

The resin known in commerce as Bombay Animé, Indian Animé, or Gum Animé, is undoubtedly derived wholly from Zanzibar, and has been referred to the above tree as its botanical source. The latest collected information on this subject was communicated in a note to the Gardener's Chronicle (1865), of which the following is the substance:—"Lieut.-Col. Playfair, H.M. Consul at Zanzibar, has recently transmitted "to the Kew Museum, in answer to a request of the late Sir W. J. Hooker, some "fine specimens of the bark of a tree with the resin in situ, together with fruits of "the same plant and various specimens of collected resin. On the bark the resin "is thickly formed in large irregular masses; the external appearance is of a dusky "grey, but its fracture is of a clear light shining yellow. The tree yielding this resin "Col. Playfair thinks is the true copal of the mainland of Zanzibar; this is an "announcement of some importance, in support of which we anxiously look for "further confirmation. Believing, however, the fruits to be what Col. Playfair sends "them for, viz., the fruits of the Zanzibar copal tree, there seems little doubt, on "comparing them with specimens in the Kew Herbarium, that they belong to

" *Hymenæa mossambicensis*, Kl. Col. Playfair says in his report to the Foreign Office " in reply to Earl Russell's inquiries respecting the supply of resin, that the value of " copal exported from Zanzibar amounted during the year 1863–4 to 163,353 dollars, " the average price being about six pounds for one Austrian dollar." These sums were made up as follows :—

United Kingdom - - -	30,030	dollars.
British India - - - -	50,044	"
Kertch - - - - - -	500	"
United States - - - -	5,000	"
Hamburg - - - - -	30,000	"
Italy - - - - - -	2,339	"

If, therefore, a revenue like this is derived from the resin of this particular plant in Zanzibar it would be well if Madagascar, where the plant likewise grows, could open up a traffic in the same commodity.

Capt. Burton says of the Zanzibar copal tree, that out of its trunk " canoes have " been formed 60 feet long, and that a single tree has sufficed for the keelson of a " brig; the average size however is about half that height, with from 5 to 6 feet girth " near the ground." The trunk " is dotted with exudations of raw gum, which is " found scattered in bits about the base;" the resin is also found in a semi-fossil state, sometimes in places overflowed by the high tides, or when sinking piles for huts, &c.

At Zanzibar the resin is sifted and cleansed from impurities before it finds its way into the hands of the foreign merchants; after this it is again cleansed by washing in chemical solutions of various degrees of strength, care being taken to thoroughly purify, but at the same time not to injure the so called " goose-skin " or the impress of the sand.

All parts of the plants of the copal-yielding species of Hymenæa are highly charged with resin, and in none is it more apparent than in the fruits of the species sent home by Col. Playfair. The little warts or verrucosities on the surface of these pods are lumps of clear colourless resin, simply covered with the thin epidermis or cuticle of the pod. To such an extent are these pods charged with resin that they burn freely upon the least approach of flame. The same property is observable in the pods of the South American species *Hymenæa courbaril*, from the trunk of which exudes very large quantities of resin, but this pod being larger and more ligneous and the resin distributed in small ducts all over the surface without penetrating the substance of the pod, does not burn so readily, nor is the quantity of resin so large. The outside of these pods much resembles in appearance the sand-fretted surface of the resin. (*Gard. Chron. Dec.* 16, 1865.)

In a letter from Zanzibar, dated 1869, Dr. Kirk writes to Dr. Hooker :—" In the " dense forest I was struck with the immense number and size of the copal trees, far " exceeding anything that I had before imagined. In no instance did I see the soft " juice flowing: where found on the tree it was invariably hard. I send a fine " specimen removed from the living tree; it will show you that now large masses, " equalling in size the fossils, are still produced, and full of insects as were those of " the ancient forests. In the dense jungles, where these trees are found the largest, " there is no under-grass to catch fire and to destroy the fallen trunks; when a tree " dies it rots until, eaten by white ants, it falls piece by piece on the ground; any " resinous masses would thus be preserved. Struck by the number of these trees, I " commenced turning up the sandy soil to the depth of 10 inches, when I soon found " small pieces of *Animi*, but not of the modern sort. There was no copal-tree in the " place where I dug; and the resin had lain there, it may be, for centuries. This led " me to examine minutely the outer skin; and I found that copal when dug up has " no trace of the goose-skin upon it. I am now satisfied that this is due to a change " in the surface of the mass after exposure, when, to a certain depth, an oxidation " takes place, or it may be a molecular change, rendering the skin more brittle than " the inner mass. When treated with a solution of caustic soda, this brittle crust " softens, and, on drying, cracks to the full depth to which the change has taken " place. After drying in the sun, the friable crust may be removed with a hair brush, " and then for the first time we see the goose-skin, characteristic of good or fossil " *Animi*. (*Journ. Linn. Soc.*)

" When cleaning copal, it may be that each merchant throws away as much as thirty pounds daily of this dusty resin brushed off the copal, and this they call sand; but it is genuine resin and, I suspect, of value, at least worth more than being cast into the sea, as is here done."

The remarks of Capt. Burton in his "Lake Regions" throw considerable light on the history of Zanzibar Copal, or Animé. He writes:—

"The copal tree* is called by the Arabs shajar el sandarús, from the Hindostani chhandarus; by the Wasawahili msandarusi; and by the Wazaramo and other maritime races mnángú. The tree still lingers on the island and the mainland of Zanzibar. It was observed at Monibosah, Saadani, Muhonyera, and Mzegero of Uzaramo, and was heard of at Bagarmoyo, Mbuamaji, and Kilwa. It is by no means, as some have supposed, a shrubby thorn; its towering bole has formed canoes 60 feet long, and a single tree has sufficed for the keelson of a brig. The average size, however, is about half that height, with from 3 to 6 feet girth near the ground; the bark is smooth, the lower branches are often within reach of a man's hand, and the tree frequently emerges from a natural ring fence of dense vegetation. The trunk is of a yellow-whitish tinge, rendering the tree conspicuous amid the dark African jungle growths; it is dotted with exudations of raw gum, which is found scattered in bits about the base; and it is infested by ants, especially by a long ginger-coloured and semi-transparent variety, called by the people maji-m'oto or 'boiling water' from its fiery bite. The copal wood is yellow tinted, and the saw collects from it large flakes; when dried and polished it darkens to a honey-brown, and being well veined it is used for the panels of doors. The small and pliable branches, freshly cut, form favourite 'bakur,' the kur bajor bastinadoing instrument of these regions; after long keeping they become brittle. The modern habitat of the tree is the alluvial sea-plain and the anciently-raised beach; though extending over the crest of the latter formation it ceases to be found at any distance beyond the landward counter slope, and it is unknown in the interior.

"The gum copal† is called by the Arabs and Hindus sandarus, by the Wasawahili sandarusi, and by the Wanyam wezi, who employ it like the people of Mexico and Yucatan as incense in incantations and medicinings, sirokko and máminúugu. This semi-fossil is not 'washed out by streams and torrents,' but 'crowed' or dug up by the coast clans and the barbarians of the maritime region. In places it is found when sinking piles for huts, and at times it is picked up in spots overflowed by the high tides. The East African seaboard, from Ras Gomani in S. lat. 3° to Ras Delgado in 10° 41', with a medium depth of 30 miles, may indeed be called the 'Copal Coast;' every part supplies more or less the gum of commerce. Even a section of this line from the mouth of the Pangani River to Ngao (Monghou) would if properly exploited suffice to supply all our present wants.

"The Arabs and Africans divide the gum into two different kinds. The raw copal (copal vert of the French market) is called sandarusiza miti, 'tree copal,' or chakazi, corrupted by the Zanzibar merchant to 'jackass' copal. This chakaji is either picked from the tree or is found, as in the island of Zanzibar, shallowly embedded in the loose soil, where it has not remained long enough to attain the phase of bitumenization. To the eye it is smoky or clouded inside; it feels soft; it becomes like putty when exposed to the action of alcohol, and it viscidizes in the solution used for washing the true copal. Little valued in European technology, it is exported to Bombay, where it is converted into an inferior varnish for carriages and palanquins, and to China, where the people have discovered, it is said, for utilising it, a process which, like the manufacture of rice paper and of Indian ink, they keep secret. The price of chakazi varies from 4 to 9 dollars per frasilah.

"The true or ripe copal, properly called sandarusi, is the produce of vast extinct forests, overthrown in former ages either by some violent action of the elements, or exuded from the roots of the tree by an abnormal action which exhausted and destroyed it. The gum buried at depths beyond atmospheric influence has, like amber and similar gum-resins, been bitumenized in all its purity, the volatile principles being fixed by moisture and by the exclusion of external air. That it is the produce of a tree is proved by the discovery of pieces of gum embedded in a touchwood which crumbles under the fingers; the 'goose skin,' which is the impress of sand or gravel,

* As the attention of the Expedition was particularly directed to the supplies of copal in East Africa, by Dr. G. Buist, LL.D., secretary to the Bombay branch of the Royal Geographical Society, many inquiries and visits to the copal diggings were made. In the early part of 1857, specimens of the soil and subsoils and of the tree itself were forwarded to the society.

† Our word copal is said to be Mexican. The gum is often confounded with gum animi, which differs from it, however, by being soluble in alcohol and by softening in the mouth. When burnt the copal exhales a faint aromatic odour. Properly speaking copal is not a gum, as it does not dissolve in water; moreover it is not affected by sun or by heat. According to some informants it may be dissolved in linseed and other oils by adding a small lump of camphor.

shows that it was buried in a soft state; and the bees, flies, gnats, and other insects which are sometimes found in it delicately preserved, seem to disprove a remote geologic antiquity. At the end of the rains it is usually carried ungarbled to Zanzibar. When garbled upon the coast it acquires an additional value of one dollar per frasilah. The Banyan embarks it on board his own boat, or pays a freight varying from 2 to 4 annas, and the usher or Government tax is 6 annas per frasilah, with half an anna for charity. About 8 annas per frasilah are deducted for 'tare and tret.' At Zanzibar, after being sifted and freed from heterogeneous matter, it is sent by the Banyan retailer to the Indian market, or sold to the foreign merchant. It is then washed in solutions of various strengths; the lye is supposed to be composed of soda and other agents for softening the water; its proportions, however, are kept a profound secret. European technologists have, it is said, vainly proposed theoretical methods for the delicate part of the operation, which is to clear the goose-skin of dirt. The Americans exported the gum uncleaned, because the operation is better performed at Salem. Of late years they have begun to prepare it at Zanzibar, like the Hamburg traders. When taken from the solution, in which from 20 to 37 per cent. is lost, the gum is washed, sun-dried for some hours, and cleaned with a hard brush, which must not, however, injure the goose skin; the dark 'eyes,' where the dirt has sunk deep, are also picked out with an iron tool. It is then carefully garbled with due regard to colour and size. There are many tints and peculiarities known only to those whose interests compel to study and observe copal, which like cotton and cashmere shawls requires years of experience. As a rule, the clear and semi-transparent are the best; then follow the numerous and almost imperceptible varieties of dull white, lemon colour, amber-yellow, rhubarb-yellow, bright red and dull red. Some specimens of this vegetable fossil appear by their dirty and blackened hue to have been subjected to the influence of fire; others again are remarkable for a tender grass green colour. According to some authorities, the gum, when long kept, has been observed to change its tinge. The sizes are fine, medium, and large, with many subdivisions; the pieces vary from the dimensions of small pebbles to 2 or 3 ounces; they have been known to weigh 5 lbs., and it is said at Salem a piece of 35 lbs. is shown. Lastly the gum is thrown broadcast into boxes, and exported from the island. The Hamburg merchants keep European coopers, who put together the cases, whose material is sent out to them. It is almost impossible to average the export of copal from Zanzibar. According to the late Lieutenant-Colonel Hamerton it varies from 800,000 to 1,200,000 lbs. per annum, of which Hamburg absorbs 150,000 lbs., and Bombay two lacs' worth. The refuse copal used formerly to reach India as 'packing,' being deemed of no value in commerce; of late years the scarcity of the supply has rendered merchants more careful. The price also is subject to incessant fluctuations, and during the last few years it has increased from 4 dol. 50 cents. to a maximum of 12 dollars per frasilah.

"According to the Arabs, the redder the soil the better is the copal. The superficies of the copal country is generally a thin coat of white sand, covering a dark and fertilising humus, the vestiges of decayed vegetation, which varies from a few inches to a foot and a half in depth. In the island of Zanzibar, which produces only the chakazi or raw copal, the subsoil is a stiff blue clay, the raised sea-beach, and the ancient habitat of the coco. It becomes greasy and adhesive, clogging the hoe in its lower bed, where it is dotted with blood-coloured fragments of ochreish earth, proving the presence of oxidising and chalybeate efficients, and with a fibrous light-red matter, apparently decayed coco-roots. At a depth of from 2 to 3 feet water oozes from the greasy walls of the pit. When digging through these foundations the gum copal occurs in the vegetable soil overlying the clayey subsoil.

"A visit to the little port of Saadani afforded different results. After crossing three miles of alluvial and maritime plain covered with a rank vegetation of spear-grass and low thorns, with occasional mimosas and tall hyphœnas, which have supplanted the coco, the traveller finds a few scattered specimens of the living tree and pits dotting the ground. The diggers, however, generally advance another mile to a distinctly formed sea-beach, marked with lateral bands of quartzose and water-rolled pebbles, and swelling gradually to 150 feet from the alluvial plain. The thin but rich vegetable covering supports a luxuriant thicket, the subsoil is red and sandy, and the colour darkens as the excavation deepens. After 3 feet fibrous matter appears, and below this, copal, dusty and comminuted, is blended with the red ochreish earth. The guides assert that they have not hit upon the subsoil of blue clay, but they never dig lower than a man's waist, and the pits are seldom more than 2 feet in depth. Though the soil is red, the copal of Saadani is little prized; being of a dull white colour it is usually designated as 'chakazi.'

"On the line inland from Bagamoyo and Kaole the copal-tree was observed at rare intervals in the forests, and the pits extended as far as Muhonyera, about 40 miles in direct distance from the coast. The produce of this country, though not first-rate, is considered far superior to that about Saadani.

"Good copal is dug in the vicinity of Mbuamaji, and the diggings are said to extend six marches inland. The Wadenkereko, a wild tribe mixed with and stretching southwards of the Wazaramo, at a distance of two days' journey from the sea, supply a mixed quality, more often white than red. The best gums are procured from Hunda and its adjacent districts. Frequent feuds with the citizens deter the wild people from venturing out of their jungles, and thus the Banyans of Mbuamaji find two small dows sufficient for the carriage of their stores. At that port the price of copal varies from 2 dol. 50 cents. to 3 dollars per frasilah.

"The banks of the Rufiji River,* especially the Northern district of Wánde, supply the finest and best of copal; it is dug by the Wawande tribe, who either carry it to Kikunya and other ports or sell it to travelling hucksters. The price in loco is from 1 dol. 50 cents. to 2 dollars per frasilah; on the coast it rises to 3 dol. 50 cents. At all these places the tariff varies with the Bombay market, and in 1858 little was exported owing to the enlistment of 'free labourers.'

"In the vicinity of Kilwa, for four marches inland, copal is dug up by the Mandandu and other tribes; owing to the facility of carriage and the comparative safety of the country it is somewhat dearer than that purchased on the banks of the Rufiji. The copal of Ngao (Monghon) and the Lindi creek is much cheaper than at Kilwa; the produce, however, is variable in quality being mostly a dull white chakazi.

"Like that of East African produce generally the exploitation of copal is careless and desultory. The diggers are of the lowest classes and hands are much wanted. Near the seaboard it is worked by the fringe of Moslem negroids called Wamrima or Coast clans; each gang has its own mtu mku or akida 'ao (Mucaddum, head man), who by distributing the stock contrives to gain more and to labour less than his followers. In the interior it is exploited by the Washenzi or heathen, who work independently of one another. When there is no blood-feud they carry it down to the coast, otherwise they must await the visits of petty retail dealers from the ports, who enter the country with ventures of 10 or 12 dollars and barter for it cloth, beads, and wire. The Kossi, south-west or rainy monsoon, is the only period of work; the Kaskazi or dry season is a dead time. The hardness of the ground is too much for the energies of the people; moreover 'kaskazi copal' gives trouble in washing on account of the sand adhering to its surface and the flakes are liable to break. As a rule the apathetic Moslem and the futile heathen will not labour whilst a pound of grain remains in their huts. The more civilised use a little jembe or hoe, an implement about as efficient as the wooden spade with which an English child makes dirt pies.

"The people of the interior 'crow' a hole about 6 inches in diameter with a pointed stick, and scrape out the loosened earth with the hand as far as the arm will reach. They desert the digging before it is exhausted; and although the labourers could each it is calculated, easily collect from 10 to 12 lbs. per diem, they prefer sleeping through the hours of heat and content themselves with as many ounces. Whenever upon the coast there is a blood feud, and these are uncommonly frequent, a drought, a famine, or a pestilence, workmen strike work, and cloth and beads are offered in vain. It is evident that the copal-mine can never be regularly and efficiently worked as long as it continues in the hands of such unworthy miners. The energy of Europeans, men of capital and purpose, settled on the seaboard with gangs of foreign workmen, would soon remedy existing evils; but they would require not only the special permission, but also the protection of the local government. And although the intensity of the competition principle amongst the Arabs has not yet emulated the ferocious rivalry of civilisation, the new settlers must expect considerable opposition from those in possession. Though the copal diggings are mostly situated beyond the jurisdiction of Zanzibar, the tract labours under all the disadvantages of a monopoly; the diwans, the heavy merchants, and the petty traders of the coast derive from it, it is supposed, profits varying from 80 to 100 per cent. Like other African produce, though almost dirt-cheap, it becomes dear by passing through many hands, and the frasilah, worth from 1 to 3 dollars in the interior, acquires a value of from 8 to 9 dollars at Zanzibar." (*Burton, Lake Regions*, p. 435.)

---

* Pedro Rezende, Secretary to the Count of Linhares ("Breve Tratado," &c. Anno 1635, quoted by Mr. Guillain, in a work before alluded to), mentions amongst the exports from Monfia (Mafiyah island) a quantity of resin which appears to be copal procured from the banks of the Rufigi River.

A writer in the Geographical Magazine also adds the following notes on copal. The Msandarusi or copal tree (*Trachylobium sp.*) according to Captain Elton is largely scattered over an extensive tract of country in the Mrima or coast region. This tract lies along both banks of the Lufiji and extends from the mountains to the sea, a belt from 30 to 35 miles wide. Throughout these limits immense quantities of the semi-fossil are dug up by the natives, and this constitutes the most lucrative commerce of the Indian settlers at the small trading ports. Farther south, beyond the Lufiji there is a break in the supply, attributed by Captain Elton, to the surrounding slave traffic, which rapidly drives legitimate commerce out of its course, but also affected by the increased difficulties of communication caused by the marshy swamps which here fringe the coast more deeply than above the Lufiji. Beyond Kilwa copal again appears, and is largely bought up in tranquil times at the numerous trading stations which dot the seaboard down to the Rovuma River.

The copal tree is a large tree. The bark is smooth; the trunk of a yellowish white tinge rendering the tree conspicuous amid the dark African jungle growths, and is dotted with exudations of raw gum, which is also found scattered in bits about the base. The forests, called kiregesi, contain many Msandarusi or copal trees, and between them, in the broad transverse glades which always intersect African woodlands, some of the finest fossil-gum is dug. This never reaches the trader, however, without a large admixture of the copal from the neighbouring trees, and the contents of the digger's basket are made up with wet sand and small stones in order to give it a little extra weight. The chief centre of supply is the Kwale district, where there are eleven stations for carrying on the copal trade. In the early morning strings of natives are seen on the paths, each party led by a few men armed with old muskets and bows and arrows, and consisting of women and lads carrying copal baskets. Except during the very dry season these arrivals take place daily, but there is no organised system of working. The Indian traders do not venture to send out parties of their own, for each village and each working is represented by a head man, and the natives are only too ready to unite against the slightest enroachments on their monopoly. The "trade union" system is here represented in its strongest form.

Below four feet no gum is found worth taking and very few diggers go beyond three feet in search of it. The trade appears to be surrounded by many difficulties. The Indian trader on the Mrima coast has many extortions to contend against, and heavy duties to pay; besides the continual haggling with the natives, competition with his neighbours, and a perpetual round of coast fevers. The local village head man extracts a ground rent from him; and he has to pay an arbitrary per-centage on his profits and twenty out of every hundred frasilahs of copal shipped as an export duty. The cost is from 3 to 5¼ dollars per frasilah of copal, varying according to the season of the year, and at Zanzibar the merchants buy at from 7 to 8 dollars. Yet in spite of all difficulties the trade prospers, and Captain Elton is of opinion that the inexhaustible supply of Copal, under a settled rule, and with systematic working, would furnish the means of supporting a far larger community than that which is now sparsely scattered along the coast.

In old days the trees would appear to have lined the shores, but the extent of the ancient forests can now only be estimated by the area of the present workings, and by the positions of the existing Msandarusi which are found away towards the foot of the low hills bordering the Mrima, and on all the terraced lands sloping down from the ridges to the present sea-beach. Msandarusi is the true or ripe copal, the produce of vast extinct forests, and buried at depths beyond atmospheric influence. The raw or tree copal is called Chakazi. It is little valued in European markets but is exported to Bombay and China. (*Gard. Chron.* 1874, p. 274.)

The specimens in the Museum collection have been derived from Bombay, Aden, and Zanzibar.

IMPORTS of COPAL into BOMBAY PRESIDENCY.

—	1869–70.	1870–1.	1871–2.	1872–3.
From	cwts.	cwts.	cwts.	cwts.
Africa, Coast	299	852	965	966
Straits, Settlements	—	—	3	—
Total	—	—	—	—

EXPORTS of COPAL from BOMBAY PRESIDENCY.

—	1869-70.	1870-1.	1871-2.	1872-3.
To	cwts.	cwts.	cwts.	cwts
United Kingdom	198	634	75	312
Suez	—	—	5	—
France	—	—	—	1
Turkey	—	—	4	4
Persian Gulf	13	—	—	17
Straits Settlements	—	—	18	12
China	—	57	1	14
	—	—	—	360
To				
Burma	—	—	—	13
Bengal Presidency	24	32	80	132
Madras Presidency	—	—	1	66
	—	—	—	—

Pistacia lentiscus, *Linn.*

Linn. Sp. 1455. *DCand. Prod.* ii. 65. *Woodv. Med. Bot.* t. 11. *Bot. Mag.* t. 1967.

Suffruticose. Leaves evergreen, abruptly pinnate, leaflets about eight, lanceolate; petiole winged. Flowers very small.

Hab.—Southern Europe, Northern Africa, and the Levant.

Roomie-mustakie, Koondur-roomie. *Hind.*
Mastic. *Eng.*
Mustika. *Arab.*
Kundir-roomee, Sakiz-roomee. *Pers.*

Only a very small quantity (if any) of the European mastic finds its way into India, nearly all being brought down from the North.

It occurs in the form of small irregular yellowish tears, brittle, of a vitreous fracture, becoming soft and ductile when chewed, having a faint agreeable odour, increased by the application of heat or friction. It contains a minute portion of volatile oil, about 90 per cent. of a resin soluble in alcohol (mastichic acid) and about 10 per cent. of a resin (masticine) insoluble in cold but soluble in hot alcohol. (*Ind. Pharm.* p. 58.)

Tournefort says that in Scio the extraction of mastic commences on the first of August. The bark is cut crosswise with huge knives. The mastic exudes and hardens partly on the stem partly on the ground. The same incisions furnish mastic towards the end of September, but in lesser quantities. The mastic which concretes on the stem is called mastic in the tear, while that which falls to the earth constitutes common mastic. (*Pereira.*) Of late years mastic has been scarce and dear in the English markets.

Pistacia cabulica, *Stocks.*

Stocks in Hook. Journ. iv. 143. *Walpers' Ann.* iv. 446.

Tree 10–20 feet high, bark tuberculate, minutely pubescent; leaves 5–7–9 leaflets, petioles from the base to the apex inconspicuously subulate, leaflets subsessile, narrow lanceolate-oblong, pointed; fruit rounded, compressed.

Hab.—Beloochistan Hills and throughout Affghanistan.

Tree called—

Kussoor. *Affghan.*

Resin called—

Mustaka, Arab, Uluk-baghdanee, Auluk-dagdadie. *Ar.*
Roomie-mustakie, Koondur-roomie. *H.*
Kinneh, Kinnoli. *P.*
Mastic. *Eng.*
Saht-i-kundiroo. *Scinde.*
Sakiz Khinjuk. *Affghan.*
Aluk-ool-unbat. *Royle.*

Pistacia Khinjuk, *Stocks.*

Stocks in Hook. Journ. iv. 143. *Walp. Ann.* iv. 446.

Tree 10–20 feet high, bark even, young parts pubescent, adult smooth; leaves 5–7–9 leaflets, petiole terete, leaflets unequilateral at the base, subfalcate, shortly petiolate, lanceolate (attenuated at the base), or ovate (base more or less rounded), attenuated at the apex into a setaceous point or rarely abrupt. Fruit obovate sub-compressed.

Hab.—Common in the mountains of Beloochistan from 4,000 to 8,000 feet. Also in Affghanistan.

Tree called—

Gwun, Gulungoor. *Beloochistan.*
Khinjuk, Shurumna. *Affghan.*

The fruit is eaten and from its seed an oil is extracted.

Resin called—

Mustaka, Arah, Uluk-baghdanee, Auluk-bagdadie. *Arab.*
? Roomie-mustakie, Koondur-roomie. *H.*
Kundiroon (Kinneh, Kinnoli), Sakiz-shereen. *Pers.*
Mastic. *Eng.*

Both these Pistachias yield a resin (according to Stocks) which is used as a succedaneum for mastic in Beloochistan, Affghanistan, Scinde, and, as he believes, in Persia also.

These trees are noticed by Griffith in his "Affghan Jounal" under the name of Khinjuk, and appear to have been regarded by him as species of Xanthoxylon (p. 351, 412). *Pistachia Khinjuk* has large yellow galls, like an old worm-eaten tamarind pod, and these are generally full of resin. These Royle mentions (*Himal. Bot.* p. 178) as Gool-i-pista and the resin as Aluk-ool-unbat. *Pistacia Cabulica* also has red excrescences like the comb of a cock, growing from the midrib on the under side of the leaf. Kœmpfer (*Amœnitates*, p. 414) mentions these trees as growing on the mountain Bunna, seven days' journey north of Bunder Abbas, and also about Shiraz, especially on a mountain near Majin one stage from Shiraz. The names he assigns to the trees are the same as those used in Beloochistan, allowing for dialect and pronunciation; one being called *Bun* or *Wun* (in Beloochistan *Gwun*), and one *Kussoodan* (in Beloochistan *Kussoor*). He mentions also the galls, and the resin, which is called Kundiroon by the Persians, or sometimes *Sakiz Shereen* (Masticha dulcis) to distinguish it from olibanum which is called *Kundir* or *Sakiz Tulkh* (Masticha amara), and from mastic itself which is called *Kundir Roomee* or *Sakiz Roomee* (Masticha Turcica). The Khinjuk by which the tree is known in Affghanistan is a slight alteration of *Wun-juk* or *Gwun-juk*. Forskal mentions this resin in his Materia Medica Kahirensis (No. 23 among the gums) as "Kuteerah Ajimee, Humrah, or Kusrooee, è Persia." I cannot say absolutely that Kœmpfer's trees are identical with the ones described from Beloochistan, but it is probable from the similarity of the names and the great resemblance which the Beloochistan flora has to that of South Persia. (*Stocks in Hook. Jour.* iv. 144.)

The mastic in the Museum Collection is from Bombay, and was probably imported there from Scinde.

Diospyros glutinosa, *Roxb.*

Roxb. Fl. Ind. i. 533. *Wi. Icon.* t. 844.

EMBRYOPTERIS GLUTINIFERA. *Roxb. Cor. Pl.* i. t. 70.

DIOSPYROS EMBRYOPTERIS. *Pers. Ench.* ii. p. 624. *DCand. Prod.* viii. 235.

Tree 25–30 feet; leaves alternate, linear-oblong pointed, glabrous, shining, short petioled; male peduncles axillary, solitary, 3–4 flowered; stamens 20; females 1 flowered, larger than the male; stamens 2–4, short; pistils 4; nut globular, size of a small apple, rusty-coloured, filled with pulpy juice, and covered with a rusty farina; seeds 8, flowers white.

Hab.—Peninsula. Travancore. Assam. Bengal.

Native synonyms—

Tumi, Tumiki, Tubiki, Tinduki, Tumika, Tumei. *Tel.*
Gaub, Gab. *H.*
Gaub, Tumika ? Gab. *B.*
Panichi-maram ? Tumbikai, Tumbika, Panichika. *Tam.*
Gab ? Sindica. *Sans.*
Timberee-gass, Timberri. *Cyn.*
Cusharatha-mara, Kusharta-mara. *Can.*
Pani-jika. *Mal.*
Panichie ? (*Travancore.*)
Gusvakendhu. *Uriya.*
Kendu. *Ass.*

The Tendoo gum and Gaup resin are probably the same substance, but of this we have no definite information. In Assam it is stated that the Kendoo tree fruit yields a resinous gum, which is used for caulking the small canoes used in Assam. The Dome fishermen colour their nets with it and thereby render them more durable in wear and tear. It commands no trade, has no marketable price, and is only prepared as required for the above purposes. In Dacca the fruit of the Gab yields a gum exclusively used in caulking boats, and the tree might be grown to any extent. (*Agri. Hort. Journ.* i. n.s. pt. i. (1867) p. xxxvii.)

The Museum specimen is a dark brown, rather earthy looking resin, with a bright resinoid fracture, not unlike some light-coloured varieties of the black dammar. A similar specimen in the reference collection is a brown resin, with a shining fracture, externally coated with a greyish white or brownish crust. It is somewhat stalactitic as if it had trickled down the tree as it was exuded. This is a different character to that given to Gaup resin in the above quotation wherein it is said to be derived from the fruit. It should be determined, therefore, whether any gum or resin exudes from the trunk of this tree, as well as more decided information obtained regarding the gummy substance derived from the fruit. The resin in the collection referred to this source was derived from Bhaugulpore; it is soluble in turpentine or benzole, forming a limpid varnish.

UNDETERMINED RESINS.

DAMMAR DAGING or Rose Dammer.—Specimen of this Dammar was sent to the London Exhibition of 1862 from Singapore, where the value is stated to be 4½ Rs. per picul. No information of its source is given, or any indication of the locality whence it was derived. Since then another specimen of the same resin was received from Johore. It is in large irregular masses, internally of a rosy brown, or bright madder brown colour, mottled with opaque streaks and irregular blotches of cream colour. Under the same name a resin was received from the Malay Peninsula, but this is transparent amber, more resembling what is termed "East India dammer" than any other resin in the collection. Rose dammer dissolves freely in turpentine or benzole, producing a slightly coloured varnish. It is really a hard copal.

ANDAMAN DAMMER.—A stalactitic mass of resin from the Andaman Islands is in the collection; it is externally striate and coated with a dirty greyish white crust, having internally a clear pitchy brown colour, and breaking with a bright vitreous fracture. It is brittle, and approaching in character pale varieties of the black dammer of Southern India, without any perceptible odour. It dissolves in a like manner to the black dammar of Travancore either in spirits of turpentine or benzole, producing a comparatively pale varnish not to be distinguished from that of the latter resin.

BORNEO DAMMER.—There is a clear amber coloured resin in the collection which it is quite possible may be referred to *Anisoptera melanoxylon* (Hook. Linn. Trans. xxiii. p. 160), but this is uncertain. Dr. Hooker states that this tree yields in Borneo and Labuan a clear yellow resin.

DAMMAR FRUTIK.—Is another Borneo resin of which no authentic specimen is in the collection. Dr. Hooker refers its source to an undetermined species of Vatica (*Linn. Trans.* xxiii. p. 161.)

DAMMAR HITAM or SALANGAN PUTIK.—This is a dark resin which is yielded in Borneo, according to Dr. Hooker, by an undetermined species of *Hopea*.

DAMMAR TRITAN or BLACK DAMMER.—Dr. Hooker observes that an undetermined species of *Hopea* (Linn. Trans. xxiii. 161) yields most of the dammer of the Malay countries, to which the above native name is assigned. This may be the source of some of the black dammers now assigned to *Canarium strictum*, and especially a fine black dammar in large stalactitic masses from Malacca in the Museum collection.

Inquirendæ.

Reviewing the foregoing observations of the present division it will be evident that there are several points on which more detailed information is desirable if anything is to be done in England for the benefit of the trade of India in copalloid resins.

I. Is the resin of the Moal of Sylhet (*Vatica lanceæfolia*, Blume) collected or produced in any quantity available for commercial purposes? And whether or not authentic specimens should be sent over for valuation and analysis? It would appear that this same tree is found and yields resin in Burmah; further confirmation of this is desirable accompanied by samples of the resin.

II. One kind of dammar batu is believed to be produed in Malacca by the tree recently described in Hooker's Flora Indica, under the name of *Shorea sericea*, Dyer.

It is desirable to know if this is confirmed, and what cream coloured resin of the Ral kind is produced in Malacca, with botanical specimens of the tree and samples of the resin, if other than the *Shorea robusta.* The sample of resin in the reference collection which most resembles "Ral" is more opaque and may possibly have a somewhat different composition. It is suspected that this is not the produce of *Shorea robusta* but of some other allied species.

III. The observation that a portion of the black dammer of the bazaars of Southern India is produced by this tree requires confirmation or refutation. Samples of resin in the collection referred to *Shorea Tumbuggaia* are of the semi-opaque cream coloured kind, resembling the "Ral," but which is therefore not at all liable of being confounded with black dammer. If any dark coloured dammer is produced by this tree samples should be sent. It will not be difficult to obtain authentic specimens of the resin of the "Googulana" from Goomsur or Cuttack, and at the same time to ascertain whether the resin of the Saul tree is also produced in the same localities. On comparing the botanical descriptions of the two species it will not be difficult to determine accurately the source of the resin.

IV. It is assumed that the greater part, if not all, the black dammer of Southern India is the produce of *Canarium strictum,* Roxb. But the question naturally arises as to what is the source of a similar resin in Assam, Tenasserim, Malacca, and Burmah. The Malacca resin appears to differ materially from the rest, whilst those of Tenasserim and Burmah are similar. From what trees are the black dammers of these several localities collected?

V. A black resin, often in very large and irregular masses, perforated irregularly in all directions with holes and passages more or less rounded, upwards of a half inch diameter, called in the lists Poon-yet or Pwai-ngyet, has been on two or three occasions received from Burmah. The resin itself is black, coated with dull greyish white and not so brittle as the black dammer of Southern India, or other black dammer from Burmah. If broken the fragments have a tendency to adhere together, and the fingers become sticky when handling fresh broken pieces. There is moreover a strong balsamic odour, resembling that of elemi in the resin, especially when freshly fractured. It is very desirable that the tree should be identified which yields this resin, and also that which affords a more brittle black dammer in which the fracture is clean and vitreous, and without any perceptible odour. The last kind is not perforated or channelled in the same way as the Poon-yet.

VI. The pitchy brown dammer from the Andaman Islands to some extent resembles the odourless black dammer. It would be an advantage to have the source of this resin also ascertained.

VII. The resin of *Canarium Bengalense,* Roxb., which is said to be so common in the bazaars of Calcutta, is required. It would be well to ensure that one sample at least should be collected from the tree in order that it may be authentic, and that its examination, analysis, and valuation may be relied upon.

VIII. Is there any evidence of the occurrence of *Dammara orientalis,* Lam., in the Malay Peninsula, or anywhere on the mainland of the Asiatic continent?

IX. What is the source of the rose dammar, "dammar daging," of Johore, and perhaps of other neighbouring places, and to what extent is it produced? Is it an article of commerce in the local markets?

X. Is any other resin known in any of the localities to which these requisitions may come, other than those enumerated, but which possess similar qualities of solubility in turpentine, ether, or benzole, so as to form a clear transparent varnish, such as that produced by the copals of Africa, or the white dammer of the Eastern Archipelago. If any such resins are known it is desirable that steps should be taken to ascertain accurately by securing botanical specimens of the leaf, flower, and fruit, to what source they may be referred, as well as the native names of the tree and of the resin, with a statement whether it is an article of local consumption, or if it can be obtained in any quantity for trade purposes.

Sect. B.—ELEMI OR SOFT RESINS.

Most of these resins are so soft and clammy to the touch as to form a very natural link between the resins of a copalline character, and the oleo-resins, which latter by parting with much of their volatile oil by exposure, will sometimes become almost solid. Some of the Elemi resins are wholly, and others in great part, soluble in rectified spirit. In many instances there is a decided, but not unpleasant odour. No product in this group produced in India is of commercial importance, the so-called Indian Elemi being derived from Manilla, although the supposed tree has been introduced into India. The home of Elemi producing trees is more especially the tropical and subtropical region of the New World.

Calophyllum inophyllum, *L.*

DCand. Prod. i. 562. *Roxb. Fl. Ind.* ii. 606. *Wi. & Arn. Prod.* 103. *Wight Illus.* i. 128. *Wight, Icon.* t. 77. *Beddome Flor. Sylv. Gen.* xxii.
CALOPHYLLUM BIUTANGOR, *Roxb. Fl. Ind.* ii. 607.

A middling sized tree; bark grey, smooth; leaves 4–8 by 3–4 in., coriaceous, shining on both surfaces; veins many, fine, petiole $\frac{1}{2}$–$1\frac{1}{4}$ in.; racemes in the upper axils, loose, 4–6 in. long, shorter than the leaves, lax, few flowered; flowers $\frac{3}{4}$ in. diameter, pure white, fragrant; pedicels slender 1–2 in.; sepals four; petals four, like the inner sepals; stamens numerous, filaments in four bundles; ovary globose, stipitate, style much exceeding the stamens, stigma peltate, lobed; fruit 1 in. diameter, globose, smooth, yellow, pulpy. (*Hook. Fl. Ind.* 273.)

Western Peninsula; from Concan and Orissa southwards. Ceylon. Eastern Peninsula; from Pegu southwards. Andaman Islands. Malay Archipelago, &c.

The native synonyms of the tree are—

Sultanchampa, Surpunka. *H.*	Oondee. Surpunka. *Dec.*
Pinnay. *Tam.*	Ponna. Poona. *Mal.*
Poona. *Tel.*	Teldomba. *Cey.*

G. Bennett (*Mag. Nat. Hist.*) says that in the South Seas a scented gum resin is produced from this tree, which at first exudes of a clear amber colour, but soon after becomes of a greenish yellow. He also adds that in Tahiti it was formerly a sacred tree, and planted in the Morai, where it was death to break a branch, or in any way to injure it. The gum resin which exudes both spontaneously and on incisions being made in the trunk, is the Tacamahaca resin of commerce, which was formerly held in high estimation as an ingredient in warm stimulating plasters, &c. The females of Tahiti use the resin as a scent, as also the fruit, which is possessed of much fragrance, and is said to yield a yellowish dye.

Birdwood remarks in his catalogue, "It is stated in books of authority that a resin " exudes from the roots of this plant, and that it is the Tacamahaca of the Isle of " Bourbon; the true East Indian Tacamahaca is produced by *C. calaba*, and *myanas* " resin is referred to the same species." On the same subject O'Shaughnessy writes that the roots (*C. inophyllum*) yield a resin deemed the same as the Bourbon Tacamahaca, the fixed oil is fragrant, and used as an external application in rheumatism.

The Pharmacopœia of India alludes to this subject in these terms, " from the bark " exudes a resinous substance which has been thought, apparently erroneously, to be " the Tacamahaca of old pharmacologists."

In the United States Dispensatory, under the head of "Tacamahac" it is said that " another variety is obtained from the East Indies, and called *Tacamahaca orientale* " or *Tacamahaca in testis.* It is supposed to be derived from the *Calophyllum inophyllum* and comes into the market in gourd shells covered with rush leaves. It " is of a pale yellow colour inclining to green, slightly translucent, soft and adhesive, " of an agreeable odour, and an aromatic bitterish taste. It is at present very rare " in commerce. The tree which yields this resin produces a drupe, about as large as " a plum, from the seeds of which 50 per cent. of a greenish yellow fixed oil is obtained " by expression, used in India for lamps, and as a local application in the itch." (*Journ. de Pharm.* 1861, p. 23.)

Guibourt describes several other varieties of Tacamahac, which, however, are little known. Among them is a soft, adhesive dark green resin, said to be procured from the *Calophyllum Tacamahaca* growing in the islands of Bourbon and Madagascar.

Without attempting to determine what may have been the source of Tacamahac, and especially of East Indian Tacamahac, there is no evidence substantially in favour of *Calophyllum inophyllum*, or any other Indian tree, and it may safely be affirmed that the Tacamahaca orientale is not obtained from India, nor is it known in Indian commerce. The resin of *Calophyllum inophyllum* must therefore be held guiltless of Tacamahac.

The resin of this tree is undoubtedly by no means a common product in India, and very rarely reaches Europe, never as an article of commerce. Only one sample in the museum collection has the appearance of being genuine.

The museum sample from Travancore is evidently incorrectly named, it being a black soluble gum, whereas the Mysore sample is a blackish resin, greenish when powdered, and may be the produce of *Calophyllum*.

Calophyllum calaba, *Jacq.* (*and other species*).

O'Shaughnessy states that this is a native of Travancore and produces true East India Tacamahaca resin. This resin is said to be rarely met with, it is yellow, translucent, adhesive, of acrid taste, and pleasant aromatic smell. The tree here indicated would appear to be an error, as it is not a native of India and consequently is not included in Hooker's Flora Indica now in course of publication. Dr. Ernst observed that in Venezuela "Incensio macho" is a name sometimes given to Tacamahaca, a resin derived from *Calophyllum calaba*, Jacq., so that the tree probably yields a resin of the Tacamahaca kind in countries where it is found. It is uncertain to what Travancore tree O'Shaughnessy alludes, but it may be the *Calophyllum Wightianum*, Wall., which extends along the mountains of the Western Coast of the Western Peninsula from the Concan to Travancore. It is figured on plate 106 of Wight's Icones, but whether any resin is collected from it or not we have no direct information. Bouton says (*Medical Plants of Mauritius*, p. 22) that *Calophyllum spurium*, Choisy, which is a synonym of *C. Wightianum*, "is found in India and Mauritius and yields a " resin acting as a vulnerary, resolutive, and anodyne."

Canarium commune, *Linn.*

Wi. & Arn. Prod. i. 175. *Roxb. Fl. Ind.* iii. p. 137.
Colophonia Mauritiana. *DCand. Prod.* ii. 179.
Bursera paniculata. *Lam.*

Tree, 50 feet; leaves unequally pinnate; leaflets 7–12 on long stalks, ovate-oblong, acute, or shortly acuminate, entire, glabrous; panicles terminal, divaricated; flowers 2–3 together, almost sessile at the extremity of the ultimate pedicels; drupe covered with a thin somewhat fleshy sarcocarp; calyx three lobed, externally silky; petals three; nut very hard, three-angled; seed solitary; flowers white.

Hab.—Peninsula. Malabar.

Native name—

Junglee-badam. *H.*

The concrete resinous exudation (Elemi) is chiefly imported into England from Manilla. It occurs in the form of a soft unctuous adhesive mass, becoming harder and more resinous by age; of a yellowish white colour, with a rather fragrant fennel-like odour, almost entirely soluble in rectified spirit. This is the so-called East India or Manilla Elemi, the probable source of which is *Canarium commune*. (*Ind. Pharm.* p. 51.) In their recent "Pharmacographia," Hanbury and Flückiger do not think that the Manila Elemi is the product of a *Canarium*.

Commercially no Elemi is derived directly or indirectly from India, and although the tree is found there, under the name of Java almond or *Junglee badam*, its resin seems to be unknown.

Ailanthus Malabarica, *DCand.*

Wight Icones, t. 1604. *Wi. & Arn. Prod.* 150.

Tree, leaves abruptly pinnated; leaflets quite entire, ovate-lanceolate, unequal sided, oblique at the base; panicles large, terminal; peduncles and calyx pubescent, petals glabrous, obovate, much longer than the calyx; samaræ oval-oblong, obtuse at both ends.

Hab.—Travancore. Malabar.

Native names of tree—

Peroomarum. *Mal.* Peru marum. *Tel.*

Native name of resin—

Mutti-pal. *Tam.*

Gum resin of a dark pitchy brown colour, homogenous, soft, running into a doughy mass which becomes dry and somewhat brittle externally after long exposure. Odour rather strong, balsamic.

In a communication to this office the following account is given of this resin:— " As commonly met with it is dark brown or grey in colour, plastic, opaque, and has " an agreeable smell. It contains much impurity. The pure resin is very soft,

" having the consistence of thick treacle, and this is doubtless the reason why it is " always mixed with fragments of wood and earth which make it more easy to handle. " The sample which I examined contained but 77 per cent. of resin, the remainder " being adulterations. Alcohol readily dissolves the resin, and on evaporation, leaves " it as a very viscous, transparent, light brown semi-liquid, which does not solidify " for many days' exposure to a steam-heat; when burnt it gives out a fragrance, and " hence it is sometimes used for incense. Its perfume is, however, inferior to that " produced by many other resins employed in the concoction of the incense employed " in Christian and heathen worship. The peculiar consistency of the resin would " enable it to substitute Venice turpentine for many purposes, though its price " (6 Rs. per 25 lbs. in the crude state) forbids an extensive employment."

Dr. Wight has stated that the bark of the tree above named is "rough and very " thick, studded with bright garnet looking grains apparently of a resinous nature " which do not dissolve either in spirit or water." These grains can scarcely be the exuding resin, which has by no means a garnet colour, and is easily soluble in spirit.

Buchanan was the first who noticed this resin and found the tree producing it in the Anamullay forests. For a long time subsequently to this the tree was lost sight of, until General Cullen found it growing in the Travancore forest, and forwarded specimens of the flowers and branches to Dr. Wight, who immediately recognised the long lost Ailanthus (*Drury*, *Useful Plants*).

The museum specimens are from Madras, Travancore, and Cochin. One sample is cream coloured and has apparently been run into bamboo.

Cassia auriculata, *L.*

W. & Arn. Prod. 290.

SENNA AURICULATA, *Roxb. Fl. Ind.* ii. 349.

Shrub; young branches, petioles, and peduncles pubescent; leaflets 8–12 pairs, with a gland between each pair, oval, obtuse, or retuse, mucronate, upper side glabrous, under slightly pubescent, racemes axillary, nearly as long as the leaves, many flowered, approximated towards the ends of the branches; pedicels compressed; sepals slightly hairy; legumes compressed, straight; flowers 3–5 together, bright yellow. (*Drury*.)

Hab.—Common in the Peninsula.

Native names are—

Thongadee, Tanghedu. *Tel.*	Turwer. *H.*
Tangayree. *Can.*	Averia. *Tam.*

This resin is held in very great estimation in a medicinal point of view, and it is so scarce that it seldom falls into the hands of any other than those chucklers who are constantly in the habit of cutting the bark of the plants for the purpose of tanning leather. These men generally sell the resin, and derive some profit thereby. (*Mysore List Exh.* 1867.)

Kunnee.

Under No. 872 in the list of Indian gums, &c. exhibited by the late Dr. Royle at the Great Exhibition of 1851, occurs *Icica resinifera* with the native names of Kunnee-gond and Jingun-ke-gond, said to have been derived from the Khera pass. It is somewhat difficult now to determine with certainty what this specimen was, and all trace of its original locality is lost. There is however no ground for believing that it was the produce of any species of *Icica*, and the only clue to its identity can be obtained from a remark in the "Illustrations" of the same authority (p. 178), where he remarks that a gum resin exudes from the bark of *Odina Wodier* which is so abundant as to form an article of commerce; "it is called in Northern India " Jhingun-ke-gond and Kenni-ke-gond. The latter name or nearly the same is " given by Dr. Ainslie as the Hindoostanee of Galbanum." From the identity of the native names in this quotation with those in the entry of the Exhibition list it may be fairly assumed that the resin was the same in both instances.

Unfortunately scarcely any other Indian product is so uncertain here as the gum of *Odina Wodier*, from time to time represented by specimens forwarded either to exhibitions or for report, and seldom on any two occasions the same substance. At one time a clear amber soluble gum of the Arabic kind bears the name of *Odina Wodier*, at another time a gum similar in appearance, but only partially soluble, containing a large per-centage of bassorin, whilst on one occasion a black pitchy resin, not in the least soluble in water and of no prepossessing appearance bore the same name. Amongst conflicting specimens it is not so easy to determine which is the true gum of *Odina*, whilst giving the remark of Dr. Royle its due weight that it is a "gum-resin," and consequently not a soluble gum. Dr. Stewart says that "it is used in cloth printing and is officinal, being given in asthma, and as a cordial "to women." O'Shaughnessy calls it a gum resin, and Drury, though calling it a "gum" says that it is beaten up with cocoa nut milk, and applied to sprains and bruises.

Baden Powell in Punjab Products (pp. 397, 416) writes of a sample of this gum being exhibited in stalactitic white semi-transparent pieces, intermingled with little bits of bark, which again appears to be some other gum confounded probably through its native name. This gentleman when in England not only expressed a strong opinion that the gum of *Odina Wodier* is a soluble gum, but had no doubt that the museum specimens referred to this source are correct. It is evident that there are two conflicting opinions as to the produce of *Odina Wodier*, one in favour of a resin, not unlike that produced by *Icica*, and the other that it is a soluble gum of the *Acacia* kind. The latter seems to us to have the advantage, and hence it remains to show what is the source of the resin which has been referred to *Odina*.

DAMMAR KEJIE from the Malay Peninsula is a resin resembling Elemi, but of which we have no information, except that it is used medicinally. It is soluble except impurities in spirits of wine.

GUTTA RUNGHAS is another resin having much the appearance of Manilla Elemi, and somewhat of the odour. It is cream coloured, firm and brittle, easily reduced to powder. It was sent from Malacca in 1862. The coloured portion is dissolved out by spirits of wine, forming a straw coloured solution, slightly sticky to the fingers. The white residue, scarcely diminished in bulk but pulverulent, is wholly soluble in spirits of turpentine.

MAL-SHAKSHI is another resin in the collection resembling elemi in odour and appearance; its locality is unknown, but it was sent to the Vienna Exhibition, it is supposed from Bombay, in 1873. It has a decided olive colour in the mass, and has a tendency to adhere to the fingers. It speedily and entirely becomes dissolved in spirits of wine, forming a pale green fluid not unlike Cajeput oil in appearance, and upon evaporation crystallized in acicular crystals. Unfortunately little reliance can be placed on the names attached to the gums, &c. sent to Vienna since the labels were in many cases transposed to the great confusion of the collection, and no special information was forwarded, sometimes not even a list.

Mr. E. Solly describes a new and peculiar Indian gum resin in the proceedings of the Committee of Commerce and Agriculture of the Royal Asiatic Society (1841, p. 144.) It states that he had lately received a sample of an Indian gum resin which had been sent from Bombay under the denomination of Gum Galbanum, but which, as it only resembled Galbanum in colour, was soon found to be a very different substance, and unsaleable. Mr. Solly was unable to make out the history of the gum resin or the name of the tree from which it was obtained; it appeared to be a new and undescribed substance. The resinous part of it when separated from gum and impurities was a pale yellow colour, very transparent, and formed when spread upon wood a clear and brilliant varnish, which, from the resin itself being soft and ductile, was free from an objection common to many varnishes, namely, of being very brittle, and apt to split off when the varnished surface is struck with any hard substance. It therefore appeared probable that it might be used advantageously mixed with some other resinous substance in the manufacture of varnishes.

No native names are given and therefore it is hardly possible to trace the above resin, or identify it with any yet known with any amount of certainty, but it is not improbable that it might have been the Mal-shakshi above alluded to, which is not at all unlike *Galbanum* in colour, although agreeing therewith in nothing else.

Group IV.—OLEO-RESINS.

Under this group it will be found convenient to subdivide the products under three subsidiary heads, which will include all the Indian representatives :—

A.—Balsams.

B.—Natural varnishes or lacquers.

C.—Turpentine and tar.

Under the first of these such oleo-resins as Cajeput oil, wood oil, storax, &c. will naturally be placed. In the second the Burmese, Sylhet, Cinghalese, Japanese, and other varnishes will be associated together. And in the last not only turpentine and tar as derived from coniferous trees, but also from teak and other trees resembling it in properties and uses.

A.—BALSAMS.

A restricted application of the term "balsam" was formerly in vogue, under which the product so classed was described as containing benzoic acid. This definition is not accepted here, on the contrary, all oleo-resins (with the exception of the acrid lacquers, or natural varnishes, and the turpentine series) are grouped together under this section. The majority of Indian balsams approximate rather to the balsam of Copaiba type than to that of balsam of Peru. The latter might be accepted perhaps as the type of true balsams, and the former of pseudo-balsams, but this subdivision need not be insisted on here.

Balsamodendron Berryi, *Arn.*

Arnot Ann. N. Hist. *Wight Illus.* p. 185. *Bedd. Fl. Syl.* 126.
Protium Gileadense, *W. & A. Prod.* 176.
Amyris Gileadensis, *Roxb. Fl. Ind.* ii. 246.
Balsomodendron Gileadense, *DCand. Prod.* ii. 76.

Tree, ultimate branches short, patent, spinescent with small very short abortive branchlets, bearing at their extremities the leaves and flowers; leaves palmately 3-foliolate; leaflets roundish obovate, obtuse, obscurely crenulate, glabrous; peduncles 1-flowered, shorter than the petioles. (*W. & A.*)

The native names of the balsam are—

Aqovoyalasamun, Dohnul-balsan. *Ar.*
Roghane-balsan. *P.*
Balsan-ka-tel. *H.*

The whole tree, says Beddome, has a grateful fragrance, and a gum resin exudes from it. This is all the information he gives on the subject. Dr. Birdwood says that this balsam is the balm of Scripture. The βαλσαμον of Theophrastus and Dioscorides, called also opobalsamum. The wood xylobalsamum and fruit carpobalsamum are also described by the ancients, and sold to the present day in Bombay. Pliny, Diodorus, and ancient authors generally considered Judæa the native country solely of the balm trees, but we now know that they are found in Arabia. They would appear to have been confined to gardens in Judea, and the vineyards of Engeddi are supposed to have been balsam groves. Diodorus gives Engeddi and the Dead Sea shore as the habitat of the trees. Calmet states that the Arabians have a tradition that the Queen of Sheba introduced them there on her visit to Solomon. They were an offering which must have been prized, for centuries later Pliny informs us that the Emperors Vespasian and Titus had the shrubs exhibited in Rome, and that the Romans were in the habit of carrying them in their triumphal processions, and also that Alexander the Great when in Judæa thought it a fair midsummer day's work to fill a concha (from ·0412 to ·1238 of a pint) with opobalsamum. Calmet derives the word balsam from baal-shemen, royal oil.

Tradition is rich in anecdote, says Fee, relative to the origin of this balsam. The Mahomedans affirm that it sprung from the blood of the slain in Mahomet's conflict with the tribe of Harb, and that the Prophet used the balsam for the resuscitation of the dead.

Dr. Ainslie writes of it that this liquid resin he had never seen in Lower India, it was, however, an article in the Ulfaz Udwiyeh, and, therefore, could not be unknown to the Moguls. This balsam is considered almost as a panacea in Egypt, where it is prescribed for bad wounds, ulcers, poisonous bites, and also in nervous and pulmonic affections.

O'Shaughnessy describes the balsam as a liquid resin of syrupy consistence, limpid, yellowish. By age it thickens and solidifies. The odour is like anise, and very penetrating; flavour aromatic, bitter, and hot; it is almost always adulterated with turpentine, olive oil, or wax. According to Vanquelin's analysis this balsam is soluble in alcohol, with the exception of a very minute residue.

It is a very fair subject for inquiry as to whether the Indian tree is identical with that of Arabia and Ethiopia, probably not, and consequently the balsam may be somewhat different.

Dipterocarpus turbinatus, *Gærtn.*

Gærtn. Fruct. iii. 51. t. 188. *Roxb. Fl. Ind.* ii. 612. *Cor. Pl.* iii. 10. t. 213. *Hook Fl. Ind.* 295.

DIPTEROCARPUS LÆVIS, *Ham. in Mem. Wern. Soc.* vi. 299, *D Cand Prod.* xvi. 607.

DIPTEROCARPUS INDICUS, *Bedd. For. Rep.* 1864–5, p. 17.

Tree, young branches compressed; leaves ovate or broadly lanceolate, entire, or sinuate-crenate, acute, base rounded, glabrous on both surfaces or slightly pubescent; stipules with dense stellate canescence, pubescent towards the apex; racemes 3–5 flowered; flowers 3 in. diam.; calyx tube obconic hoary-puberulous, in fruit 1 in. diam., subsphæroidal, mouth contracted; unenlarged lobes deltoid-ovate; petals pinkish white, not pubescent; fruit wings 5 in. by 1 in. linear-oblanceolate, obtuse, 3 nerved.

Hab.—Eastern Bengal and Eastern Peninsula, from Chittagong and Pegu to Singapore.

Synonyms—

Garjan-ka-tel. *H.*	Hora-tel. *Cing.*
Gorjon-tail. *B.*	Kaninsi, Kanyen-si, Kanyeen. *Burm.*

The class of substances called wood oils, writes Lieut. Hawkes, form the connecting link between the oils and resins of the vegetable kingdom. They consist of a volatile oil holding in solution a resin, and are generally classed under the head of balsams. They are obtained from various trees of the order *Dipterocarpeæ*, some of which also yield the dammers of commerce. The mode of extracting the oil is similar to that adopted for obtaining the solid resin. It is best described by a contributor to the Madras Exhibition of 1855, who says, "About the end of the dry season, that is, in "March or April, several deep incisions are made with an axe into the trunk of the "tree, and a good sized piece scooped out, into these holes fire is placed and kept "burning till the oil begins to run, which is received into a bamboo and allowed to "run slowly drop by drop. The oil when freshly obtained from the tree and allowed "to rest separates into two layers, the upper consisting of a clear chestnut coloured "liquid balsam, and the lower being a flocculent deposit of the more solid resin of a "light ash colour. They are much used as natural varnishes for indoor work, but are "very brittle and require constant renewal. Perhaps the admixture of a certain "proportion of drying oil would remedy this defect. They are said to be used also "in the manufacture of lithographic inks. These oils are chiefly imported from our "Burmese possessions, and the islands in the Straits of Malacca, and are usually "known by the name of the district from whence they are brought."

Roxburgh says that to procure the balsam a large notch is cut into the trunk of the tree, near the earth (say about 30 inches from the ground) where a fire is kept up until the wound is charred, soon after which the liquid begins to ooze out. A small gutter is cut in the wood to conduct the liquid into a vessel placed to receive it. The average produce of the best trees during the season is said to be sometimes 40 gallons. It is found necessary every three or four weeks to cut off the old charred surfaces and burn it afresh; in large healthy trees abounding in balsam they even cut a second notch in some other part of the tree, and char it as the first. These operations are performed during the months of November, December, January, and February. Should any of the trees appear sickly the following season, one or more years' respite is given them. (*Fl. Ind.* ii. 613.)

Wood oil, as imported from Moulmein, is after filtration a transparent dark brown liquid, of somewhat greater consistence than olive oil, a sp. gr. of ·964 and an odour and taste like copaiba, though perhaps hardly so strong. One part of it treated with two parts of alcohol, sp. gr. ·796, is dissolved with the exception of a minute quantity of darkish flocculent matter, which subsides upon repose. But its most curious property is that exhibited when it is heated in a corked vial to about 266° Fahr. Thus treated it becomes slightly turbid, and so gelatinous that the vial may be inverted, even while hot, without its contents being displaced, and on cooling the solidification is still more complete. Gentle warmth and agitation restore to a great extent its fluidity but solidification is again produced upon the liquid being heated to 266°. Copaiba displays no such phenomenon.

According to Dr. O'Shaughnessy, when wood oil is heated in a retort a yellowish white, crystallisable solid substance having many of the properties of benzoic acid sublimes into the upper part of the vessel, to the extent of about one per cent. of the wood oil taken. It is true that when wood oil is heated a scanty opaque white sublimate condenses in the cooler part of the vessel, adds Mr. Hanbury, but this appears to arise from the condensation of a little water among the minute drops of essential oil, since it is not produced if the wood oil has been previously agitated with some fragments of dried chloride of calcium. (*Pharm. Journ.* xv. 321.)

In appearance this balsam is dark coloured and turbid. Its turbidity is due to a greenish resinous matter, held in suspension, which is, however, easily separable, either by filtration or deposition, leaving a brown transparent liquid of sp. gr. 0·970. When the latter fluid is submitted to a careful distillation it yields—

Essential oil	65	per cent.
Resin	34	,,
Acetic acid and water	1	,,
	100	

I found that the essential oil in its various reactions with potassium, iodine, nitric acid, &c., and moreover in taste, exactly corresponds with those presented by pure essence of copaiba. The resin left by distillation of the balsam, either with or without water, is, if deprived of the whole of the essential oil, extremely hard. Its entire solubility in coal naptha proves the absence of any of the soft resin which exists in most of the copaiba of commerce. This hard resin (copaivic acid) being most probably the active principle of balsam of copaiba, I am induced to think its quantity and purity in the one I have examined is indicative of its superior value as a medicament. The dark colour of the balsam may perhaps limit its employment, but the large quantity of copaivic acid it contains renders it valuable, as the latter may be made available by heating the filtered balsam to the boiling point with a small quantity of caustic potash or soda lye of sp. gr. 1·34 and separating the resinate of potash or soda from the essential oil. The alkaline resinate may then be dissolved in water, giving a colourless solution, or the balsam may be treated with magnesia to form the ordinary copaiba pill. (*Lowe in Pharm. Journ.* xiv. 65.)

Lieut. Hawkes says that the wood oils are much used as natural varnishes for indoor work but are very brittle, and require constant renewal. Dr. Wight speaks from experience of the value of Gurjun oil mixed with dammar in preventing the white ants from attacking timber. There appears to be one great disadvantage in the use of wood oil as a varnish in this country, that it dries so slowly and has so little body. Samples recently submitted to London drug brokers for report were declared to be without definite value in this market, as they are comparatively unknown and not in demand.

Dipterocarpus incanus, *Roxb.*

Roxb. Hort. Beng. 42. *Flor. Ind.* ii. 614. *W. & A. Prod.* 84. *DCand. Prod.* xvi. 611. *Hook. Fl. Ind.* 298.

Young parts and compressed branches softly hairy pubescent; leaves ovate obtuse, 6 by 4½ in., thinly pubescent on both surfaces, margins ciliate, lateral nerves 12 pairs, pubescent beneath; petiole 2 in., softly pubescent; calyx tube (in flower) softly pubescent, strongly winged. (*Dyer.*)

Hab.—Chittagong and Pegu.

Roxburgh says Gurjun is the vernacular name in Chittagong, where the tree grows to a great size, and is said to furnish the largest proportion of the best sort of wood oil or balsam.

Dipterocarpus alatus, *Roxb.*

Roxb. Hort. Beng. 42. *Flor. Ind.* ii. 614. *Hook. Fl. Ind.* 298. *DCand. Prod.* xvi. 611.

DIPTEROCARPUS COSTATUS, *Gærtn. Fruct.* iii. 50, t. 187.

DIPTEROCARPUS GONOPTERUS, *Turcz in Mosc. Bull.* 1868, 576. *DCand. Prod.* xvi. 612.

OLEOXYLON BALSAMIFERUM, *Wall. Cat.* p. 157.

Young branches compressed, pubescent. Leaves ovate or elliptic acute, $5\frac{1}{2}$ by 8 in. shining above, more or less pubescent beneath, margins ciliate, lateral nerves 15 pairs, petiole $1\frac{1}{2}$ in., softly pubescent, hairy; stipules pilose, racemes about 7 flowered; calyx tube in fruit globose, 1 in. diameter, smooth; fruit wings (calyx lobes) 4–5 by 1 in. linear ovate or spathulate, obtuse, 3-nerved, unenlarged lobes orbicular.

Hab.—Chittagong, Burmah, Tenasserim, and the Andaman Islands.

This is one of the trees which is said to yield wood oil. Roxburgh says that it is the wood oil tree of Pegu.

Dipterocarpus tuberculatus, *Roxb.*

Roxb. Fl. Ind. ii. 614. *DCand. Prod.* xvi. 614. *Hook. Fl. Ind.* 297.

Leaves cordiform, acute, base cordate or truncate, margins sinuate-crenate, undulate; calyx-tube in fruit pentangular-costate in its upper part, mouth contracted; flowers large, rose-coloured.

Hab.—Chittagong. Burma.

Native names of the tree:—

Eng, Enben, *Burm.*

A wood oil, under the name of Eng, said to be the produce of *Dipterocarpus tuberculatus,* was sent from Burmah (May 1874) for valuation and report.

Chloroxylon swietenia, *W. & A.*

Wi. and Arn. Prod. i. 123.

SWIETENIA CHLOROXYLON, *Roxb. Fl. Ind.* ii. 400. *Roxb. Cor. Pl.* i. t. 64.

Tree: leaves abruptly pinnate; leaflets pale-coloured, small, numerous, alternate or nearly opposite, unequal sided; calyx short, five partite; petals five, shortly unguiculate; panicles terminal, branched, capsule oblong, three celled, three valved, dehiscing from the apex, septifragal; seeds about four in each cell, extending upwards into a wing; flowers small, greenish white. (*Drury.*)

Hab.—Circars. Mountainous districts of the Peninsula.

Native names of the tree:—

Kodawa-pursh, Moodudad, Vummaray, Moodooda, Vum-maay, Kodawah-porsh. *Tam.*
Billoo, Billuda. *Tel.*
Boorooch-gaha. *Cey.*
Hulda, Billoo. *Bombay.*

All the information we possess of this product is to the effect that the tree yields a wood-oil, but the extent of produce or use is unrecorded, and there is no specimen in the collection.

Dryobalanops camphora, *Coleb.*

Colebr. Asiat. Res. xii. 537 (1818). *Hook. Comp. Bot. Mag.* i. 253. *Hook. Journ. Bot.* iv. (1852) 36.

SHOREA CAMPHORIFERA, *Roxb.? Fl. Ind.* ii. 617.

Tree: leaves ovate, acuminate, quite smooth; petiole short; stipules in pairs; perianth one-leaved, five parted; capsule superior, ovate, woody, finely streaked with longitudinal furrows, one celled, three valved; wings large, foliaceous, spathulate, rigid; seed solitary.

Hab.—Sumatra. Borneo.

Kapur Barus. *Malay.*

Camphor oil, the peculiar juice of the tree, exudes from its fissures and cavities, and is carefully collected. The oil is so fine that a paper penetrated by it, and held near a flame, catches fire immediately, and burns till all the oil is consumed. (*Hook. Journ.* 1852, p. 35.)

The oil seems to consist of a very volatile essential oil, holding in solution a resin, which on a few days' exposure to the air is left in a syrupy state. It also yielded a small quantity of crystallised camphor on distillation with a very rough extempore apparatus. It is used as a popular remedy for rheumatism, being rubbed on the affected part. It is also a fragrant, quickly drying, and well-bodied varnish, for which purpose the writer used it largely. It requires rubbing until dry, like French polish. China is as yet (1852) the sole market, where it is used principally for embalming. (*Hook. Journ. Bot.* 1852, p. 204.)

Both oil and camphor are found in the heart of the tree occupying a vacuum, which, in others, is frequently filled with pitch; but it does not extend to the whole length, on the contrary, they are found in small portions of a foot and a foot and a half long, at certain distances. The method of extracting the oil is merely by making a deep incision with a billiong or Malay axe in the tree, about 14 or 18 feet from the ground, till near the heart, where a deeper incision is made with a small aperture; and the oil, if any in the tree, immediately gushes out, and is received in bamboos, or any other utensil better approved of. In this manner a party proceeds through the woods wounding the camphor trees till they attain their object. (*Asiat. Resear.* vol. xii. p. 539.)

Camphor oil sometimes makes its appearance in the London markets, but only at intervals and rarely.

Hardwickia pinnata, *Roxb.*

Roxb. Fl. Ind. ii. 425. *Bedd. Fl. Sylv.* 255.

Large tree, leaves pinnate; leaflets 5–6 alternate, glabrous, ovate, obtuse, stipules minute; panicles terminal, shorter than the leaves; flowers minute on pedicels; legume indehiscent, obovate, coriaceous, almost woody; seed pendulous from the apex.

Hab.—S. Travancore Ghats. S. Canara.

Native names of tree:—

Matâyen Samprâni. *Tam.*

The liquid balsam, like Copaiba, sold at Trevandrum, and the produce of a tree from these Ghats (S. Canara) is extracted from the *Hardwickia pinnata.* (*Beddome Flora Sylv.* xciv.)

The tree yields a dark red balsam, almost exactly like the balsam of Copaiva, and with all the same properties. The tree is called in Travancore Matayen Samprani (which means fool's incense), and the balsam is extracted and used medicinally; to obtain it a deep notch is made into the heart of this tree, and after a time it begins to flow. Mr. Broughton reports of it that "this substance appears on examination to consist of " a solution of certain chemically different resins in an essential oil, and is in fact an " oleo resin." Like the wood oils from the different species of *Dipterocarpus*, it greatly resembles both in composition and properties the Copaiva balsam, though it lacks the transparency and light yellow colour of the latter. It is nearly entirely soluble in ammonia, but does not produce a clear solution. The essential oil has the same composition as that from the Copaiva balsam. It boils at a temperature of 225° C. It rotates the plane of polarisation to the left, but in a different degree to that found with the oil from Copaiva. This essential oil occurs in different amounts in the balsam, and more abundantly in the fresher collected specimens; these are quite fluid, but other specimens are almost semi-solid, doubtless owing to the evaporation and oxidation of the oil. The oil is best obtained by prolonged distillation of the balsam with water. By this means I have obtained from an apparently old specimen of balsam 25 per cent. of oil, and in the most recently collected specimen I have obtained over 40 per cent. I have made many attempts, but have not obtained crystals of Copaiva acid from the balsam. The solid resins are of an acid character, but the balsam does not solidify so strongly as that of Copaiva after being heated with Magnesia. The oil can be separated from the balsam by Adler's process, but it is obtained in a very impure and coloured state. There appears little doubt that this balsam could effectually substitute Copaiva balsam in medicine. But the appearance of the specimens that I have received is greatly inferior to the latter, and could not under present circumstances compete with the Brazilian balsam in the European markets. (*Beddome Fl. Syl.*)

Liquidamber Altingia, *Blume.*

Blume Fl. Jav. p. 8. *Miq. Fl. Ind. Bat.* i. p. 836.
ALTINGIA EXCELSA. *Noron in Verh. Bot.* v. p. 1.
SEDGWICKIA CERASIFOLIA. *Griff. Asiat. Res.* xix. p. 98, t. 15.

Large tree; trunk thick, sulcate below; bark cinereous, whitish, even or verruculose, internally reddish brown, yielding from cuts or wounds a honey-like balsam; leaves alternate, ovate or elliptic, oblong, obtusely acuminate, glandulose, serrate, smooth, shining, pallid beneath; amenta in crowded panicles.

Hab.—Sunda Archipelago. Cochin China. New Guinea. Java. Tenasserim.

Native names of the resin:—

Meaahe-sayelah. *Ar.*
Asle-lubni. *P.*
Silaris. *H.*
Neri-arishippal. *Tam.*
Shila-rasam. *Tel.*
Selaras. *Guz.*
Nantayu, Nan-ta-youk. *Burm.*
Rasamala. *Jav.*

This is a large forest tree of Java, common also in some parts of the Tenasserim Provinces, especially at Paulow, about 30 miles from Mergui. It yields a fragrant balsamic fluid, of which two varieties are met with in Burmah; one pellucid, fragrant, and of a light yellowish colour, obtained by simple incision in the bark; the other thick, dark, opaque, of a highly terebinthinate odour, obtained by boring holes in the stem, and the application of fire around the trunk of the tree. Analysed by Marquart, it was found to contain a volatile oil somewhat like Styrol, and a substance similar to styracine, but their composition he found to be different. (*Ind. Pharm.* p. 88.)

Some writers have regarded this resin as the liquid storax of commerce, but this has been shown to be an error.

Liquidambar orientale, *Miller.*

Hook. Ic. Pl. vol. xi. *Boiss. Fl. Orien.* ii. 819.

LIQUIDAMBAR IMBERBE. *Ait. Kew.* iii. 365.

PLATANUS ORIENTALIS. *Pococke Trav.* ii. pl. 89.

Tree; leaves palmate 5–7 lobed, quite smooth, lobes acute or obtuse, terminal sometimes trilobate, fruit-bearing peduncles nodding.

Hab.—Anatolia, &c.

Names of the balsam:—

Liquid storax. *Eng.*
Buchuri-jag. *Turk.*

This is not an Indian product, but being included in the Pharmacopœia of India, and met with in Indian commerce, has a place here.

The oleo-resin is thus described in the Pharmacopœia:—" A balsam obtained from " the bark. It occurs in the form of a semi-transparent, brownish-yellow, semi-fluid " resin, of the consistence of thick honey, with a strong agreeable fragrance and " aromatic taste. Heated in a test tube on the vapour bath it becomes more liquid, " but gives off no moisture. Boiled with solution of bichromate of potash and sul- " phuric acid it evolves the odour of hydride of benzoyle. It contains a volatile oil, " a crystallisable principle, and cinnamic acid."

The most exhaustive inquiry was made by Mr. Danl. Hanbury as to the sources of liquid storax, and the result of his investigations were published in the Pharmaceutical Journal (March 1857), in which he arrives at the conclusion that the liquid storax of the present day was not the *Storax* of the ancients, and that it is derived from *Liquidamber orientale.* From these communications the following information was obtained:—

The localities of the tree are south-west of Asia Minor, forests in the district of Sighala, near Melasso; forests near Moughla, and near Giova and Ulla in the gulf of Giova; also near Marmorizza and Isgen-gak opposite Rhodes.

Mr. Maltass passed through a dense forest of liquidambar between the village of Caponisi and the town of Moughla on the 7th or 8th May 1851; he describes it as consisting of trees resembling the plane, but evidently of a different species, the leaf being smaller, and each tree far denser in foliage than the plane usually is. " I also " observed," says he, " that most of the larger trees had the (outer) bark stripped off " from the trunk and the inner bark scraped off. I gathered some of the fruit and " leaves, and proceeded on my journey towards Moughlia, my road lying for upwards " of an hour through this beautiful forest. I observed that the trees were from 20 to " 30 feet in height, but whenever there was a break in the forest and the trees had " sufficient air and space they were of larger growth, many of them being 40 feet " high, but more especially in the immediate vicinity of streams of water. My guide " assured me that in some places in the forest in the direction of Melasso he had seen " some of these trees 60 feet in height. He could not tell me the name of the tree, " but stated that an oil was produced from it called *Buchar*, and that the trees were " mutilated to obtain it."

In June and July the outer bark is stripped off on one side of the tree and (according to Lieut. Campbell) made into bundles and reserved for the purposes of fumigation. The inner bark is then scraped off with a semi-circular or sickle-shaped knife, and thrown into pits until a sufficient quantity has been collected. Mr. Maltass states that it is then packed into strong horse-hair bags and subjected to pressure in a wooden lever press. Upon removal from the press hot water is thrown over the bags and they are pressed a second time, after which the greater portion of the resin will have been extracted.

Lieut. Campbell's account is a little different; he says the inner bark is boiled in water over a brisk fire, upon which the resinous part comes to the surface and is skimmed off. The boiled bark is next put into hair sacks and pressed, boiling water being added to assist in the extraction of the resin, or, as it is termed, *yagh* (*i.e.* oil).

Dr. McCraith says that the storax collectors, who are chiefly a tribe of wandering Turcomans, called *Yuruks*, are armed with a triangular iron scraper with which they scrape off, together with the juice of the tree, a certain quantity of bark, which they collect in leathern pouches suspended to their belts. When a sufficient quantity has been obtained it is boiled in a large copper and the separated liquid resin is run into barrels. The residual bark is placed in hair cloth and pressed in a rude press, the extracted resin being added to the general mass.

The product obtained by the processes here described is the grey opaque semi-fluid resin well known as liquid storax. The bark from which the liquid storax has been extracted is emptied out of the bags and exposed in the sun to dry, after which it is shipped to the Greek and Turkish islands, and to many towns in Turkey where it is much esteemed for the purpose of fumigation, although since the disappearance of the plague its employment has greatly diminished. This is the substance known to pharmacologists as Cortex Thymiamatis or storax bark, as is proved by abundant specimens sent by Mr. Maltass.

Lieut. Campbell states that the quantity of liquid storax annually extracted amounts to 20,000 okes (500 cwt., from the districts of Giova and Ulla, and 13,000 okes (325 cwt.) from those of Marmorizza and Isgengak. It is exported in casks to Constantinople, Smyrna, Syra, and Alexandria. Some is also packed with a certain proportion of water in goat skins and sent either by boats or overland to Smyrna, where it is transferred to casks and shipped mostly to Trieste.

In a report of the external commerce of Bombay to which my attention was directed by my friend Dr. Royle, the term Rose Malloes is applied to a drug enumerated under the category of imports by sea into the port of Bombay. The recurrence of this strange name, which Petiver's account of liquid storax had made familiar, struck me as very curious, and I wrote to Bombay for a sample of the drug so called. This was kindly procured for me at the Bombay Custom House by Dr. Carter and proved to be our ordinary liquid storax. It is imported chiefly from the Red Sea, which it doubtless reaches from Alexandria, to which port I found it was shipped from Rhodes. Here then is the explanation of Petiver's statement of the drug being carried southward from Suez—it being in fact on its way to India. (*Pharm. Journ.* Feb. 1857.)

Melaleuca minor, *Smith.*

Smith, in Rees Cyclop. *DCand. Prod.* 212.
MELALEUCA VIRIDIFLORA. *Bl. Bijdr.* p. 1,099.
MELALEUCA LANCIFOLIA. *Turcz. Bull. Soc. Mosc.* xx. 164.

Tree; leaves alternate, elliptic-lanceolate, acute or slightly obtuse, straight or somewhat falcate 5-7 nerved; spike interrupted, and, as well as calyces, rather villous.

Hab.—Malayan Peninsula and Archipelago, and Northern Australia.

The native synonyms for Cajeput oil are—

Kai-puti-ka-tel, Kai-buti-ka-tel. *H.*
Kaiyap-pudai-tailam. *Tam.*
Kayaputi-tail. *B.*
Kayputi-nu-tel. *Guz.*

The oil distilled from the leaves is a very mobile transparent fluid of a fine pale bluish-green colour. It has a strong agreeable odour, and aromatic camphoraceous taste, leaving a sensation of coldness in the mouth. (*Pharm. Ind.* p. 90.)

Rumphius states that the leaves are gathered on a warm day, and placed in a sack where they become hot and damp. They are then macerated in water, and left to ferment for a night, and afterwards submitted to distillation. Two sacks full of the leaves yield scarcely three drachms of the oil, which is limpid, pellucid, and volatile. Lesson has described the method of obtaining the oil at Bouron one of the Molucca Islands. The leaves, he says, are gathered in the latter end of September, and put into the cucurbit of a copper alembic, surmounted by a neck, terminated by a capital without a refrigeratory, and a sufficient quantity of water is then added. By distillation this liquid is made to traverse a worm immersed in a hogshead filled with water, and is collected in a vessel; the oil which floats is very light and of an herbaceous green colour, which is owing to chlorophylle, or perhaps a somewhat different resinous principle. By rectification it becomes colourless.

Cajeput oil is usually imported in green glass bottles (in appearance similar to long necked beer bottles). Its colour is green, the tint being that of a strong solution of chloride of copper. It is transparent, limpid, of a strong penetrating smell, resembling the combined odour of camphor, rosemary, and cardamom, and of an aromatic camphoraceous taste, succeeded by a sensation of coolness, like that caused by oil of peppermint. In the mass the odour is disagreeable, but in small quantity, as when rubbed on the hand, is much more fragrant. An apparently pure sample has a spec. grav. of 0·925. Dr. Thomson says the spec. grav. varies from 0·914 to 0·9274, while Mr. Brande states it to be 0·980. Oil of Cajeput is soluble in alcohol. Its boiling point is 343°. When carefully distilled with water the first portion of the oil which passes over is very light and quite colourless, but toward the end of the process a heavier and greenish oil distils over. (*Pereira, Mat. Med.*)

Mr. Crawford in his "History of the Indian Archipelago" states that there are three varieties of this tree, native to Amboyna, Java, and Borneo. The bark of the largest, which is a mountain tree, grows in continuous forests, and the bark of it yields the material with which the natives of the Moluccas caulk their ships. The smaller, which yields the Kyapootre or Cajeput oil, thrives near the sea coast, and has got the name from its colour, kayu-puti, which signifies white wood, and hence Rumphius gave it the name of arbor alba minor. Besides its current name, it is known in Malay countries, says Dr. Ainslie, by other terms, as *galam*, *daun*, kibsjil, &c., in Ternate by *cajule*, in Amboyna by *kilam*, *ilam*, and *elam*, and in Ceram by that of *sakelan*.

In forwarding plants for cultivation in India from Malacca, Dr. Maingay writes, " They form the first growth in low lying marshes after the heavy forest has been " removed, and as these marshes are by no means so malarious as might be expected, " I ascribe it entirely to the melaleuca growth, for the trees abounding in resin in " their leaves act upon the malarious emanations in a similar manner to the Stone " Pine of the Landes upon similar emanations in that district, and to a great extent " render them innocuous. Upon bruising the Melaleuca leaves, you at once perceive " the agreeable aromatic odour. I think, therefore, that this quality would render " them valuable as an addition to the list of trees useful for planting in marshy " pestilential spots near our Indian cantonments and elsewhere, provided the Indian " cold weather is not found too severe for them. The wood itself is highly durable " under water, and the bark, mixed with dammer, &c., is extensively used by the " natives to caulk their boats, and might be used for making coarse paper similar to " that made from the mulberry bark in Japan." (*Agri. Hort. Journ.*, Aug. 1867.)

Mesua ferrea, *L.*

W. & Arn. Prod. i. 102. *Wi. Icon.* t. 117. *Roxb. Fl. Ind.* ii. 605.

Tree, 40 feet, sepals 4, unequal; petals 4, alternate with the sepals; leaves oblong, lanceolate, acuminated, glaucous beneath, upper side shining, midrib and margins coloured; flowers stalked, axillary, large, white, fragrant; fruit about the size of a small apple; 1 celled, 1–4 seeded.

Hab.—Courtallum.

The native names of the tree are—

Nag-keshur. *B.*
Nagkesar. *H.*
Gangau. *Burm.*
Na-gaha. *Cyn.*
Irul-maram, Irool-maram? *Tam.*
Chikati-manu? Naga-kesarachettu, Suvarnam. *Tel.*
Kesaramu-naga-sara, Kinjalkamu. *Sans.*
Nag-champa. *Mahr.*
Behetta-champagam. *Mal.*
Nahor. *Ass.*
Nagishvoro. *Uriya.*

Mr. Brownlow remarks that the resin of the Nagessor (*Garciniaceæ*) would probably make a good substitute for Canada Balsam, as it has great viscidity, strength, and clearness, and probably high refractive power. (*Journ. Agri. Hort. Soc.* xiii. p. 346.) Round the base, or rather at the bottom of the tender fruits of *Mesua ferrea*, a tenacious and glutinous resin exudes with a sharp aromatic smell. (*Drury.*)

The above is the only information that has come to our knowledge of this balsam, of which no specimen exists in the Museum Collection. It is presumed that Mr. Brownlow's remarks apply to the *Mesua*.

Mohal Balsam.

Of this substance all that we know is the allusion made to it in the Journal quoted below. The name is insufficient for the identification of the tree. "Mohal is " a jungle tree of Cachar, from which a balsam exudes resembling in odour 'capivi.' " When the tree flowers, *i.e.* in April and May, the whole jungle is filled with its " fragrance." (*Proc. A. H. Soc.*, xii. p. xxxiv.)

Mr. Brownlow says every part of the tree has the peculiar scent of the resin which resembles that of balsam capivi. The resin of the Mohal, instead of acting as a preservative, hastens the destruction of the timber by attracting borers. (*Agri. Hort. Journ.*, xiii. 352.)

B.—NATURAL VARNISHES.

The Burmese lacquer or *Theet-tsee* may be accepted as the type of what is comprehended by this group. Substances, mostly of an acrid nature, exuded by species of *Rhus*, *Semecarpus*, and their botanical allies, occur in most warm countries, so that Burmah, China, Japan, Ceylon and others have their own peculiar varnish tree. Besides these principal sources, other closely allied species will yield, perhaps in less quantities, similar substances, so that all together a very natural group may be constituted of products having a like function, for which the above title has been selected, with the present definition of what the term is intended to convey.

Melanorrhœa usitatissima, *Wall.*

Wall. Pl. As. Rar. i. p. 10, pl. ii. 12.

Large tree; leaves obovate, obtuse, entire, coriaceous, veins numerous, reticulated, petioles short; fruit coriaceous, indehiscent, ovate, depressed or somewhat kidney-shaped, 1-celled, 1-seeded.

Hab.—Munipur. Burma.

Native names—

Kheu. *H.*
Tisi, Sissi, Theet-tsee, Zit-si. *Burm.*

Every part of the tree abounds with a thick viscid greyish terebinthinate fluid, which soon assumes a black colour on exposure to the air. This is the Black Varnish or *Thit-tsi* of the Burmese, by whom it is employed extensively, not only in the arts, but as an anthelmintic. (*Pharm. Ind.* 60.)

Wallich obtained what information he could respecting this tree and its products, and was of opinion that the same tree as was found by him in Burma extended to the N.E. frontiers of Silhet and Tippera, and was identical with the Kheu or varnish tree of that district. Capt. Grant affirmed that the tree grew in great abundance at Kubba, forming large forests in conjunction with the two staple trees of Continental India, the Saul and Teak, especially the former. Numbers of the gigantic wood oil tree (*Dipterocarpus*) were also found in company with it. The size varied, but in general it attains very large dimensions. Capt. Grant speaks of trees having clear stems of 42 feet to the first branch, with a circumference near the ground of 13 feet; and he mentions that they are known to attain a much greater size. All the individuals grow in the same manner, that is, they reach a great height before throwing out any branches.

In 1812, M. R. Smith communicated some particulars of this tree and its produce to Mr. H. Colebrooke. He writes:—"I have discovered a sort of varnish, which I " consider as the identical one made use of by the Chinese in their eastern and north- " eastern provinces. It is procurable in great quantities from Munipur, where it is " used for paying river crafts, and for varnishing vessels destined to contain liquids, " such as oil, ghee, milk, honey, or water. The drug is conveyed to Silhet for sale by " the merchants, who come down annually with horses and other objects of trade. " The tree which yields it grows to an amazing size. I am informed that it attains " one hundred cubits in height, and twenty cubits in circumference, and even more. " It forms extensive forests, which commence at a distance of three days' journey from " the capital, and stretch in a northerly and easterly direction towards China for many " miles." Dr. Wallich says that there cannot be any doubt that the tree described by Mr. Smith, and that found by Capt. Grant, are identical with the Theet-tsee or varnish tree of the Burmese. It follows hence that the tree has a very wide geographical range, extending from Munipur (lat. 25° N., long. 94° E.) to Tavoy (lat. 14°, long. 97°). The valley of Kubbu, which has been ascertained by actual survey made by Lieut. Pemberton to be only 500 feet above the plains of India, is distant 200 miles from the nearest sea shore. The tree there attains its greatest size, and becomes smaller as it approaches the sea on the coast of Tenasserim, where it grows in comparatively low situations.

Dr. Wallich also states that from his own knowledge the natives of the countries where the tree is indigenous never experience any injurious effects from handling its juices; it is strangers only that are sometimes affected by it, especially Europeans. He adds that he was acquainted with instances of the latter kind where it produced extensive erysipelatous swellings, attended with pain and fever, but not of long duration.

In the neighbourhood of Prome a considerable quantity of varnish is extracted from the tree, but very little is obtained at Martaban, owing to the poverty of the soil, and partly also to the circumstance of there being none of the people in that part whose business it is to perform the process. This latter is very simple; short joints of a thin sort of bamboo, sharpened at one end like a writing pen, and shut up at the other, are inserted in a slanting direction into wounds made through the bark of the trunk and principal boughs, and left there for 24 or 48 hours, after which they are removed, and their contents, which rarely exceed a quarter of an ounce, emptied into a basket made of bamboo or rattan previously varnished over. As many as a hundred bamboos are sometimes seen sticking into a single trunk during the collecting season, which lasts as long as the tree is destitute of leaves, namely from January until April, and they are renewed as long as the juice will flow. A good tree is reckoned to produce from 1½, to 2, 3, and even 4 viss annually (a viss being equal to about 3½ lbs. avoir.). In its pure state it is sold at Prome at the rate of one tical, or 2*s.* 6*d.*, the viss. At Martaban, where everything was dear at the time, the drug was retailed a 2 Rs. per viss. It was of an inferior quality, and mixed with Sesamum oil, an adulteration often practised.

The extensive use to which this varnish is applied, indicates that it must be a very cheap commodity. Almost every article of household furniture destined to contain either solid or liquid food is lacquered by means of it. At a village close to Pagam on the Irawaddi, called Gnaunee, where this sort of manufacture is carried on very extensively, and to great perfection, I endeavoured to obtain some information relating to the precise mode of lacquering; but I could learn nothing further than this, that the article to be varnished must first be prepared with a coating of pounded calcined bones, after which the varnish is laid on thinly, either in its pure state or variously coloured by means of red or other figments. I was told that the most essential, as well as difficult part of the operation consists in the process of drying, which must be effected in a very slow and gradual manner, for which purpose the articles are placed in damp and cool subterraneous vaults, where they are kept for several months until the varnish has become perfectly dry. Another object for which the drug is extensively employed is as a size or glue in the process of gilding, nothing more being required than to besmear the surface thinly with the varnish, and then immediately to apply the gold leaf. If it is considered how very extensively that art is practised by the Burma nation, it being their most frequent acts of devotion and piety to contribute to the gilding of their numerous religious edifices and idols, it will be evident that a great quantity of the drug must be consumed for that purpose alone. Finally, the beautiful Pali writing of the religious order of the Burmas on ivory, palm leaves, or metal, is entirely done with this varnish in its native and pure state. (*Wallich.*)

This varnish is thick at the ordinary temperature of this country, of a dull leaden grey colour, but the surface, wherever it comes in contact with the air, assumes in a very short time a shining black appearance. Alcohol, spirits of turpentine, and benzole equally dissolve and combine with it, rendering it more fluid. It may also be diluted with gold size, which tends to improve its drying properties and intensify its colour, whilst turpentine has a tendency to render it brown. At present it is reported by manufacturers to have no value in the English market as it is unknown, has no special application, and is so long in drying.

Holigarna longifolia, *Roxb.*

Roxb. Fl. Ind. ii. 80. *Bedd. Fl. Sylv.* 167.

Tree, 60 feet; leaves alternate, cuneate, oblong or acute; petioles usually with a soft incurved, thorn-like, deciduous process on each side about the middle; panicles terminal and axillary; styles recurved; calyx 5-toothed; petals five, oblong, spreading;

stamens five, shorter than the corolla ; nut ovate, with a fleshy pericarp ; flowers small, whitish.

Hab.—Travancore. Concans. Chittagong,

The native names are—

Kagira. *Can.*
Hoolgeree, *Mahr.*
Cattu-Tajeru. *Mal.*

This tree is very common about the western ghats of this Presidency (Madras) from Canara down to Cape Comorin, and is also found in the plains and close to the sea. It also occurs in Bombay and Bengal, but is absent from Ceylon. A very black acrid juice is extracted from the trunk and from the fruit rind, which is used by painters, and as a black varnish. (*Beddome.*)

Semecarpus anacardium, *L.*

Wi. & Arn. Prod. i. 168. *Wi. Icon.* 558. *Roxb. Cor. Pl.* i. t. 12. *Roxb. Fl. Ind.* ii. 83.

Tree, 50 feet; leaves entire, cuneate, obovate, rounded at the apex, whitish beneath but not downy; calyx flat, 5-cleft; petals five, sessile, spreading; flowers panicled, terminal, branched; fruit sessile, cordate-ovate, with a slight notch on one side under the apex; flowers small, green.

Hab.—Concans. Coromandel. Courtallum. Guzerat. Bengal. Travancore.

The native names of the tree are—

Bheela, Belawina, Bhelaman, Bhelawan. *H.*
Shayng-cottay. *Tam.*
Neela-jeedie, Nellajidi, Jeedighenzadoo, Bhallataki, Bhallatamu, Tummedæ-mamidi. *Tel.*
Kampira. *Mal.*
Bhilawa. *Dec.*
Gheru. *Can.*
Kiri-badulla. *Cey.*
Chaibin. *Pegu.*
Biboo. *Bombay.*

The pericarp of the marking nut tree yields a resinous juice which, according to Dr. Birdwood, is the black varnish of Sylhet.

Semecarpus Travancorica, *Bedd.*

Bedd. Fl. Sylv. pl. 232.

Large tree; leaves glabrous, oblong, rounded at apex, occasionally lanceolate, terminating in a fine point, very coriaceous, shining, paler beneath, 1 foot long by 5–6 in. broad, veins and reticulated veinlets prominent, particularly beneath; panicles axillary and terminal, about as long as the leaves, spreading and compound; flowers 5–6 mereous; calyx slightly puberulous and ciliate, male with a large turbinate rudiment of an ovary; drupe obliquely oblong, rounded at the apex, about 1 inch long; hypocarp broad and short.

Hab.—Tinnevelly. Travancore.

Native name—

Natu-shengote. *Tam.*

Very common in the moist forests on the Tinnevelly and Travancore mountains (1,000–3,000 feet); it abounds with the same caustic black juice as *S. anacardium.*

Semecarpus, *Sp.*

The black varnish of Ceylon, writes W. C. Oondaatje, is produced from a species of *Semecarpus*. This gum resin is equal to the black varnish of China for the purpose of lacquering. It exudes spontaneously from the stem and branches, and may also be obtained by making incisions in the bark. The resin is hard, breaks with a smooth shining fracture, burns with a bright flame, melts in the fire, is soluble in turpentine, insoluble in water, and adheres strongly to wood and metal. The fresh juice is very acrid, inflaming the skin, and producing pustules.

To a saturated solution of *Vateria* resin in oil of turpentine add by degrees small pieces of the black resin, put it into a bottle and shake it well until the whole is dissolved, strain, and then apply it to wood or metal.

It belongs to the same natural family of plants as the varnish tree of China and Japan, and possesses the same acrid properties when applied to the skin in a fresh state as it exudes from the bark. The resin exudes from natural fissures of the bark, and at first white, becomes afterwards black by exposure to the sun, hardening into masses of different sizes. The juice also drops on the ground around the tree forming flattened pieces of resin. Mr. Ondaatjee thus accounts for his attention being called to this tree in 1852:—"A soldier was cutting fire-wood in the neighbourhood of "Badulla, and among others he felled this tree. The juice spurted out at each cut "of the axe on his fore arm; he returned home and washed himself; the next day he "felt an itching in the arm, in the evening it was swollen, red, and painful; next "morning he was unable to wear his jacket, and went to hospital; a pustular eruption, "not unlike that produced by tartar emetic ointment, having broken out in the places "where the juice had touched the skin." (*Notes of Badulla by W. C. Ondaatje*, p. 45.)

Odina wodier, *Roxb.*

W. & Arn. Prod. i. 171. *Wi. Icon.* i. t. 60. *Roxb. Fl. Ind.* ii. 293. *Royle, Illus.* t. 31, 82.

Large tree; leaves alternate, about the ends of the branches, unequally pinnated; leaflets 3–4 pair opposite, almost sessile, oblong-obovate, acuminated, glabrous, entire, paler below; calyx shortly four-lobed, segments rounded, petals five, oblong spreading; drupe uniform, very hard, one-celled; seeds solitary, of the same shape as the nut; racemes terminal, fascicled; flowers small, greenish yellowish externally purplish. (*Drury*.)

Hab.—Coromandel mountains. Bengal. Travancore.

Native names—

Goompena-karra, Gumpena-chettu, Goompina, Dumpini, Gumpini, Goompana-chettu, Goompana-karra, Goompinee. *Tel.*
Ude, Othe-marum, Anni-carra, Ooday-maram, Annaikarai-marum, Odda-marrom, Ody-maram. *Tam.*
Shimti. *Can.*
Nabhay, Hnan-bai, Na-bai. *Burm.*
Hik-gass, Hig-gass. *Cyn.*
Mageer. *Mahr.*
Dampara. (*Godavery forests.*)
Kemal, Kyamal, Jingan, Dila, Kamlai, Kambal, *N.W.P.*

Dr. Birdwood remarks in his Bombay Products that this tree yields a varnish. The gum or resinous products of this tree are so unsatisfactorily known, as will be observed from other references made thereto in course of this report, that it is most desirable that some efforts should be made to ascertain positively what is the character of its exudations.

Buchanania latifolia, *W. & A.*

W. & Arn. Prod. i. 169. *Roxb. Fl. Ind.* ii. 385.

Tree 30 feet; leaves alternate, entire, broadly oval or obovate, obtuse; calyx small, obtusely 5-cleft; petals five, sessile, recurved; branches of the panicles hirsute, terminal, and axillary, with the flowers crowded, assuming the appearance of a corymb at the tops of the branches; fruit a drupe, with slightly fleshy-red sarcocarp; nut very hard, 2-valved, 1-celled; flowers small, greenish-white.

Hab.—Mountains of Coromandel and Malabar. Belgaum Forests. Mysore.

Native names are—

Chironji, Pujal? Piar-cheronji, Charooli, Dhan, Chirauli. *H.*
Chinna-moral. Chara-chettu, Chara-pappu, Chara-namidi-chettu? Charu-mamidi, China-moralli, Jaru-mamidi, Sarapuppoo. *Tel.*
Nuskul, Noas-kool. *Can.*
Kat-maa, Kaat-mango, Moreda. *Tam.*
Lumbo, Thit-sai? *Burm.*
Piyala? *B.*
Pia-sal. *Guz.*
Char. *Mahr.*
Chara. *Sans.*
Charo, Chāru, Bhalleah. *Uriya.*
Sarapappoo, Chara, Cheronjee-kaghar-sarai.

This tree, writes Dr. Birdwood, is said to yield a varnish. Drury also says that "an oil exudes from the cut bark used in lamps, and would probably serve as an "excellent varnish," but it is not clear whether he alludes to this species or to *Buchanania angustifolia*. It is by no means improbable that both may yield a similar substance, in common with other trees of the same natural order.

Rhus succedanea, *DC.*

DCand. Prod. ii. 68. *Roxb. Fl. Ind.* ii. 98.

Leaves pinnate, 5–7 pair, smooth; petioles round and smooth; leaflets oblong, lanceolate, acuminate, shining, reticulated with veins beneath, of the same colour; drupe the size of a pea, obliquely reniform.

Hab.—Japan. Kumaon. Nepal.

Native names :—

Titar, Tatri, Choklú, Holáshi, Hala, Hali, Kakkrin, Hulashing. *H.*

Rhus vernicifera, *DC.*

DCand. Prod. ii. 68.

Leaves pinnate, 5-6 pair; petiole wingless, and as well as the branches furnished with a soft woolly covering; leaflets ovate acuminate, quite entire, nearly smooth above, velvety pubescent beneath.

Hab.—Japan. Nepal. Kumaon. Gurhwal.

Native names :—

Gudambal, Rikhali, Arkhar, Arkol, Urkur, Hulasa, Rikhul. *H.*

Both these species of *Rhus* yield a black varnish, especially the latter, which is said to be the source of the Black Varnish of Japan. There is no indication that the produce of either is obtained in the Himalayas. Dr. Stewart includes both the species in his Punjab Plants with the remark that both are said to yield a varnish in Bombay. This latter observation, however, is undoubtedly an error. If the trees are found in Bombay at all, it would be only rarely as cultivated plants.

C.—TURPENTINE AND TAR.

Under this heading are included the resinous products of Coniferous trees, which are at first semi-fluid, although ultimately some of them become, by evaporation of their more volatile elements, firm and solid. Allied substances, such as rosin and tar, which are the result of distillation are also included. In a few instances substances very similar in their nature may be derived from trees belonging to other natural orders, and these would find a place here by having regard rather to the character of the substance itself tnan the botanical affinities of the plant from which it is derived.

Pinus excelsa, *Wallich.*

Gordon's Pinctum, p. 222.

Leaves in fives, very long, three edged, very glaucous on the inner faces, bluish green, and rounded on the outer one; from 6-8 in. long, very slender, and mostly drooping sheaths, short at first, but soon rolling up, and finally falling off. Branches in regular whorls and spreading, those near the bottom reflected, while the upper ones are more or less ascending; branchlets slender, long, and spreading; male flowers in dense clusters. Cones solitary or sometimes two or three together round the leading shoots, of a cylindrical or somewhat conical shape, from 6-9 in. long, and two inches broad near the base, tapering towards the point, and with a footstalk nearly one inch long; when young of a pea-green colour and somewhat erect, but when fully grown completely pendulous, and of a pale brown colour, full of resinous matter in the shape of transparent drops; scales thickened at the ends, but without any extended or curled points, loosely imbricated, oval, blunt pointed, thin, smooth, and nearly all of a size, being one inch and a quarter long, and about an inch in breadth. Seeds rather small, with a wing one inch and a quarter long. (*Gordon.*)

Hab.—Nepal. Himalayas.

The native names are—

Cheeltoo, Cheetoo, Cheeo, Cheela, Kair, Kail, Lim, Leem, Lamshing, Raisulla, Raesulla, Cheel, Kaal, Biar. *Him.*
Lumshing. *Bhootea.*
Kuel-cheel. *N. W. P.*
Tong-schi. *Chin.*
Kuel of *Simoor and Gurwhal.*
Yari. *Cash.*

This is clearly one of the Himalayan Conifers which is capable of affording turpentine and tar and probably does so to a certain extent. Dr. J. L. Stewart says "its resin is mentioned in a Kangra Report, but I cannot find that it is largely "produced or collected anywhere."

Pinus longifolia, *Roxb.*

Gordon's Pinetum, p. 200.

Leaves in threes, very slender, three edged, of a bright, glossy green, finely serrated on the edges, and rather pendulous, or curved backwards on the young trees, 12–14 in. long, thickly set on the gross branches, particularly towards the ends and upper parts of the plant; sheaths 1¼ in. long, permanent; male flowers produced in long close clusters of many together at the ends of the branches; round at first, but elongated as they open. Cones either singly or in clusters, varying from 3 to 5 in number, in regular whorls, 5 in. long, and 2½–3 in. circumference near the base, more or less ovate, very smooth, glossy, and hard. Scales much thickened at the ends and with a large thick hooked beak 1¼ in. wide in the larger ones, but more recurved and smaller towards the base and full of resinous matter. Seeds large, with rather a long narrow wing 1½ in. long. (*Gordon.*)

Hab.—Himalayas from Bootan to Affghan.

Dhonur-ratinaj. Dhonur-ratiyanaj. *Ar.*	Gandha-baroje-ke-tel. *H.*
Roghane-ratiyanah. *P.*	Gandah-feroze-ka-tel. *Dec.*

A very fine turpentine is obtained as exudation from incisions made on the trunk. This is chiefly valued by the natives for its rosin, and as the latter is only obtained by exposing the turpentine to heat, the oil, the more valuable product, is dissipated to procure the rosin, but by adopting a very simple still the resin is obtained as good as ever for the purposes of the natives, while the oil was found to be of very superior quality. (*Agra Ex. Cat.* 1867.)

In the Calcutta Journal for April 1842, it was stated that a sample of this turpentine obtained from Major Carter afforded 75 per cent. of rosin and 10 per cent. of rectified oil of turpentine.

Dr. Stewart states that the resin of this tree is more largely collected and used than that of any other of the Himalayan Conifers. It is procured by incision, which is detrimental and ultimately destructive to the tree. Purified and unpurified it is used internally, and as a plaster, and is sometimes employed in coating timber which is to be exposed to water. The oil of turpentine is extracted from it, but probably the practice was introduced by Europeans, as has been that of extracting tar from the wood by destructive distillation. (*Punjab Plants*, p. 227.)

The natives of the Himalayas prepare tar in a simple way from fragments of the wood. The dry chips are put into a large earthen pot with a narrow neck containing about 10 seers, and in the bottom four or five small holes one fifth of an inch diameter are drilled. The pot so filled is luted over with stiff wet mud on the top and sides. A hole being dug in the ground, a smaller pot holding three quarters of a seer is placed in it, and the larger one on the top; the joint being luted, and the surrounding space filled up with earth, a heap of cow-dung bratties is piled over the whole, and as much more as is required to keep up the fire for eight or nine hours. The residue of each pot gives five chittaks of tar and one seer of charcoal; four men will easily make 2½ maunds or nine large pots full of tar in the month, and the cost will be 21 rupees = 3 annas 8 pie per seer. The product appears to be equal to the tar imported from Europe which is prepared on a larger scale. Mr. J. D. Smithe adopted a modification of this plan at the Madhopore workshops. On the large pot holding 12 seers and filled with chips he placed a smaller one inverted, luting the joint and upper surface with stiff mud five inches thick; these vessels thus prepared are put on the top of a third, which we may call the receiver, and, as in the former method, the joint being luted, the whole is covered with fuel and a fire lighted. Four to eight hours are necessary to extract all the tar. After the fire has been extinguished and the vessels have cooled, the ashes should be raked out until the under vessel or receiver is visible; the large pot should then be carefully lifted off with a thick cloth in the hands. As the layer of mud is essential for the preservation of the vessels, time and fuel will be saved if by careful management the coating is not broken; each time it is renewed a considerable expenditure of fuel takes place. Care is necessary in lifting the large pot to prevent lute or ashes falling into the receiver which contains the tar. The charcoal should then be taken out of the large vessel, and the tar out of the receiver, when they are ready to be charged again as at first. Common bazaar pots may be used, and with good management they may be fired 10 or 12 times; the economical working of the tar factory very much depends upon care and attention. The pots may be worked in a row 1⅛ foot apart, by this arrangement there will be a great saving of fuel. The average produce, according to

Mr. Smithe is as follows :—One seer of wood, yields 2-6th chittaks of tar, and 4–3rd chittaks of charcoal, giving 6–9th chittaks as the produce of each seer of wood put into the pot, or 48·1 per cent. To procure a seer of tar, 6 seers 4 chittaks of fresh chips are required for charging a pot, and 2 maunds 6 seers and 9 chittaks of chips for fuel. The estimated cost is about one anna per seer, which however seems to be too low.

Finally it is stated that the tar produced by the above process from the chips of *Cedrus Deodara*, *Pinus longifolia*, or *Pinus excelsa* is of a superior description equal to Swedish tar. It is a mixture of resin and oil of turpentine, more or less blackened by the admixture of empyreumatic products; it thickens after exposure to the atmosphere, and may be used for coating boats; it is valuable as a preservative for all the woodwork of dams, regulating bridges, and railway sleepers, also for telegraph posts and wooden fencing.

Pinus Gerardiana, *Wall.*

Gordon Pinetum, 195. *Carrière Conif.* p. 433. *Royle Illust.* 32, pl. 85, f. 2.

Leaves in threes, stiff, three-edged, stout, and bluntly terminating in a short point, from 3–5 in. long, of a bluish green colour, and glaucous when young; sheaths short at first, and composed of dry reddish brown scales, but soon rolling up and falling off; branches ascending, lower ones spreading; branchlets short, rather slender and confused; cones ovate, oblong, or somewhat cylindrical, widest at the base, from 6–8 in. long, 12–14 in. circumference at the base, of a bluish colour; scales thick, blunt, much recurved and spiny at the points; seeds nearly an inch long, cylindrical, almost wingless, pointed at both ends of a dark brown colour, and agreeable to eat. (*Gordon.*)

Hab.—Himalayas.

The following are the native synonyms of the tree :—

Konecha, Kolecha, Chulghozeh, Julghozeh, Newr, Kuminche, Koneeunchee, Neoza, Ri, Rhee, Shungtee. *Him.*

This tree is very resinous, and is generally reckoned the best of all for torches and fuel. Major Longden says that the Kanáwaris do not use its resin as it gets too hard, but he extracted excellent tar from the wood by destructive distillation (*Punjab Plants*, p. 226). Gordon says the Neoza pine affords abundance of fine turpentine and the cones exude a copious white resin.

Pinus Massoniana, *Sieb. & Zucc.*

Carrière Conif. p. 487. *Lamb. Pinet.* ed. 2, 1, p. 16, t. 8. *Sieb. & Zucc. Fl. Jap.* t. 113. *Gordon Pinetum Suppl.* 60.

Leaves in twos, needle-shaped, stiff, straight, acute-pointed, and rough at the edges, 4–6 in. long, convex on the outer side, concave on the inner, somewhat glaucous, sheaths rather short, formed of several broad scales, fringed or jagged at the ends; buds covered with imbricated, non-resinous scales; branches rather long, spreading, covered with an ashy-grey bark; branchlets rather smooth and slender; cones solitary or sub-vertical, very numerous on old trees, 2–2½ in. long, conical, rounded at the base, tapering to the point, on short sub-reflexed footstalks; scales small, woody, linear-oblong, slightly thickened on the upper or exposed part, obliquely diamond shaped, closely imbricated, with a slender elevated median line, having in its centre a little caducous prickle; seeds small, with a rusty brown membranous wing three times the length of the seeds. (*Gordon.*)

Hab.—Japan, Burma.

Native name,—

Tinyooben. *Burm.*

Crude turpentine derived from this tree in conjunction with *Pinus Khasyana* was received from British Burma for report in 1872. The market rate is stated to be 8 annas per viss, or about fourpence per lb. It was reported on as follows :—Having carefully examined the sample of rough or crude turpentine and found the quality to be fine and equal to any ever imported from United States where the distillation for this market and that of the continent (Bordeaux excepted) is carried on. "Crude" has ceased to be an article of import. There was not any in 1872, 1871, 1870, or 1869, whereas 20 years ago it was a great trade. The nominal value of sample before us would be 13*s*. 6*d*. per cwt. It will be seen at once by comparison of these figures that if the market rate is correctly stated at 4*d*. per lb., or from 35*s*. to 37*s*. per cwt. in Burma, and the nominal value in the London market but little exceeds one-third of that amount, it cannot be expected to enter into British commerce, nor would it be remunerative for export at a greater cost than one-fourth of its present Burmese rate. Spirits of turpentine at its current rate of 26*s*. would still be 10*s*. below the Burmese price for "crude," indeed the latter would appear to be far too high for any competition with the American produce.

Pinus Khasyana, *Brandis.*

Hab.—Burmah.

Turpentine collected partly from this species and partly from *Pinus Massoniana* was received in 1872 from British Burmah for report. The result will be found under the latter species. *Tinyooben* appears to be the native name applied to both species.

Pinus Latteri, *Mason.*

Mason, Burmah, 2nd ed. p. 545.

Tree 50–60 feet high, 1½–2 feet diameter; sheaths of the leaves arranged spirally, tubular, membranous, six lines long, leaves two from each sheath, equal, from 7 to 8 inches long, acute with a sharp point, convex on the back, slightly scabrous, with eight rows, in pairs, of very minute thorns which produce a striated appearance, hollow on the under surface, serrated; cones ovate-conical, nearly four inches long; scales rhomboid, unarmed. (*Mason.*)

Hab.—Amherst.

Native name:—

Hten-roo. *Burm.*

The wood, Mason says, appears to contain more resinous matter than any other species of conifer I ever saw, and large quantities of both pitch and tar might be manufactured in the forests if a remunerative price could be obtained for the article.

Abies Smithiana, *Loudon.*

Gordon Pinetum, p. 12. *Loud. Arbor.* iv. 2317.
PICEA MORINDA, *Link. Carrière Coniferes*, p. 340.
PINUS MORINDA *of Gardens.*
PINUS KHUTROW, *Royle Illus.* pl. 84, f. 1.

Leaves singly at nearly equal distances around the shoots, mostly four-sided, somewhat curved, from 1½ to 2 in. long, scattered, and with excessive sharp points; branches spread out horizontal, those nearest the bottom somewhat bent downwards, lateral ones very numerous, slender, and drooping; cones pendulous, when fully matured, from 4–6 in. long and 2½ in. broad, ovate-oblong, or nearly cylindrical, with very even brown obovate rounded scales, coloured with a glaucous bloom when young, the young cones are at first upright, but soon become pendulous; reeds very small, nearly black, wings rather small. (*Gordon.*)

Hab.—Bootan to Kafristan at from 7,000 to 12,000 feet.

Native names of the tree:—

Chilrow, Khurok, Morinda, Row, Koodrow, Koondrow, Khut-row, Raisulla, Bhooj-rai, Bung-rai, Rai, Ryung, Kundrou, Ragha. *Him.*
Tee. *Bhootea.*

The late Dr. Royle remarked many years ago that a very excellent resin was afforded by this conifer, and Dr. Stewart states that the resin of this tree is mentioned in Kangra, but it is doubtful if it is largely produced or used anywhere.

IMPORTS of ROSIN into BOMBAY PRESIDENCY.

—	1871–2.	1872–3.
From	cwts.	cwts.
United Kingdom	398	284
Mauritius	137	—
Africa, Coast	585	750
Red Sea	—	36
Aden	317	48
Persian Gulf	163	788
Straits Settlements	125	1
Australia	—	10
Total	1,725	1,917
From		
Madras Presidency	4	—
Scinde	—	89

EXPORTS of ROSIN from BOMBAY PRESIDENCY.

—	1871–2.	1872–3.
To	cwts.	cwts.
Bengal Presidency	2,316	3,820
Madras Presidency	22	—
Scinde	2	16
Africa, Coast	79	—
Red Sea	54	—
Aden	1	—
Total	—	—

Cedrus Deodara, *Loudon.*

PINUS DEODARA, *Roxb. Fl. Ind.* iii. 651. *D Cand. Prod.* xvi. p. 408.

Large tree, coma pyramidal, large, branches verticillate, lower ones somewhat pendulous, upper spreading, all pendulous at the apex; leaves spreading or pendulous at the top of the shortened branchlets, somewhat 30-fasciculately collected, shortish, straight, stiffish, somewhat quadrangular, sides slightly compressed, green, bluntly mucronate; male aments solitary, erect, oblong, acute; antheriferous bracts stalked, ovate above, rounded, and denticulate at the apex; cones solitary, erect on a short branchlet, or on a 2-cleft branchlet twin, oval, or oval-oblong, very obtuse, not umbilicate; scales numerous, imbricated, somewhat woody; bracts small, much shorter than the scale; nuts obovate, narrowed at the base, shorter than the obovate-triangular wing. (*Drury.*)

Hab.—Himalayas.

Native name of the oleo-resin:—

Kelon-ka-tel. *H.*

This Conifer yields a coarse, very fluid kind of turpentine, which is held in much esteem by the natives as an application to ulcers and skin diseases. (*Ind. Pharm.*) The wood is used for the production of tar as well as that of *Pinus longifolia* and *Pinus excelsa*.

Tectona grandis, *Roxb.*

Roxb. Cor. Pl. i. t. 6. *Flor. Ind.* i. 600.

Large tree, ash coloured scaly bark; young shoots 4 sided, channelled; leaves opposite, oval, scabrous above, whitish and downy beneath; panicles terminal, large, cross-armed, divisions dichotomous, with a sessile fertile flower in each cleft, the whole covered with a coloured farina; peduncles quadrangular, sides deeply channelled; flowers numerous, small, white; calyx and corolla 5–6 cleft, stamens often 6; ovary round, hairy, 4 celled; cells 1 seeded, nut very hard. (*Drury.*)

Hab.—Malabar. Canara. Travancore. Goojerat. Tenasserim. Burma.

The native synonyms are—

Segoon. *H.*	Tekka. *Cey.*
Segoon. *B.*	Jati. *My.*
Thaikoo, Thaikoo-marum. *Tam.*	Jaadi. *Can.*
Teka. *Tel.*	Sagwan. *Bombay.*
Tekka, Theka. *Mal.*	

Wood tar from this tree was submitted to Mr. Geo. E. Evans, analytical chemist, for examination and was thus reported upon :—" I have examined the wood tar you " sent me and I find that it contains all the ingredients to be found in coal tar but " in different proportions relatively to each other. I am of opinion that if used in " every way in which coal tar is made available, its effects would be much less " permanent than the former, particularly if exposed to the action of the atmosphere, " but this could only be proved by actual experiment, and by noticing its effects for " a length of time. I am convinced that it might be rendered much more valuable " by concentration, say, by exposure to the sun's rays for a short time in large " evaporating pans, as it would then part with a large amount of watery vapour, " which it contains in a free state." (*Agri. Hort. Journ.* vol. xi. pp. xlviii.)

R. A. Sterndale of Seonee remarks of this substance that teak tar is not made and sold in very large quantities by the natives here, for it is used by them only medicinally, I am therefore unable to give you the district rate of sale, but from recent experiments the cost is as follows :—In the first place the wood, that which has been cut about three months is the best, if too fresh the tar is thinner; about 20 seers of the above wood will yield one seer of tar, to extract which from one to two maunds of cowdung fuel (which is always used) is required, this costs about two annas a maund; allowing the full quantity of fuel, which would be four annas, say one anna for the wood (which is over the price) and two annas daily hire to a man to attend the distilling, the maximum cost of one seer of tar is seven annas. If a larger quantity is made of course it would be cheaper, as one man could attend to several distilleries. I should say 4 annas per seer is a fair average. From three to six seers according to locality can be obtained here. Yesterday I measured off a cubic foot of good teak and had it distilled, the product was about one seer of tar, the fuel one and a half maund of dried cowdung. (*Journ. Agri. Hort.* xi. (1860), lxv.)

Sethia Indica, *W. & A.*

W. & Arn. Prod. i. 106. *Wi. Illus.* i. t. 48.

ERYTHROXYLON MONOGYNUM, *Roxb. Cor.* i. t. 88. *Fl. Ind.* ii. 449.

Small tree; petals five; leaves alternate, obovate or oblong, obtuse, cuneate at the base, feather-nerved, reticulated with veins, under side pale, pedicels axillary 1–3, about twice as long as the petiole, 1-flowered; calyx 5-lobed, styles combined nearly to the apex, longer than the stamens; drupes oblong, triangular, three-celled, two of the cells small, abortive and without seeds; flowers small, greenish. (*Drury.*)

Hab.—Circars. Travancore mountains. Mysore. Malabar. Ceylon.

Native names—

Devadaru, Tevadanum, Semmanathy, Devadaram, Sima-natti, Sembu-linga-maram ? Sembu-linja-maram ? Thavadarum. *Tam.*	Adavi-gerenta. *Tel.* Deodaru. *Dec.*

Mr. W. C. Ondaatje writes of this oleo-resin, which is called Dummele in Ceylon :— " My attention was first drawn to it by having seen it used by some moormen who " are chiefly employed in fishing on the extensive back waters of Putlam. They apply " it to dhonies, canoes, &c. as it effectually preserves the wood. The mode adopted " by the moormen to extract the resin is simple, and quite in the rude primitive style " in which things are done by the natives. They pack into a common earthen pot or " chatty having a narrow mouth pieces of the wood and invert it over another one " buried in the ground. The upper one containing the wood is plastered with clay " and cowdung to which a fire is applied, until by the smothered combustion to which " this process subjects the wood the tar flows into the lower vessel in the shape of a " viscid tenaceous semi-fluid of a brownish black colour. It is soluble in ether, " alcohol, turpentine, and I believe contains creosote. It is desirable to draw attention " to the properties of this tar as a preservative of wood. It may be easily obtained at " Putlam for a very moderate price in quantities sufficient for experiment. At present " the article is prepared only in proportion to the demand for it, which is very " limited."

APPENDIX A.

On the Genus Boswellia, *with Descriptions and Figures of three new Species. By* GEORGE BIRDWOOD, *M.D., Edinburgh.*

Linn. Trans., xxvii., *p.* III. (1869.)

I offer here the descriptions of three new species of *Boswellia*, natives of the Soumali country. The characters of one of them make it, I believe, necessary that the characters hitherto given of the genus *Boswellia* should be reconsidered; and hence the first part of the title which I have, very reluctantly, given to this paper. Another of these plants, all of which yield frankincense, yields, I believe, the bulk of the olibanum of commerce. And I believe that the discovery of these plants settles at last the controversy as to the countries and trees which produce it which has gone on for ages concerning frankincense.

The offering of incense on altars, and in cups and closed censers, is represented in painting and in sculpture on the monuments of Egypt and Assyria; but although incense, as a rule, implies frankincense, and these representations might with more or less plausibility be interpreted by the Jewish ritual, they merely record the contemporaneous use of incense. We are expressly told by Herodotus that frankincense was excluded from the balsamic substances used in the preparation of the mummies of the Egyptians; but although this might perhaps have been because of its being sacred to the service of their gods, still the first undoubted record of the use of frankincense, and the first mention of it—as yet known—is in the

BIBLE.—"Stacte, and onycha, and galbanum, with pure frankincense" were the "sweet spices" of which the "pure and holy perfume," or "confection," of divine prescription was made "after the "art of the apothecary," which was offered every morning and evening on the "Altar of Incense" or "Golden Altar," set in the "Holy Place" between the "Golden Candlestick" and "the Table of Shewbread," before the "Holy of Holies."[1] The priest took a censer, "full of burning coals of "fire," from off "the Altar of Burnt Offering" or "Brazen Altar," and his "hands full of the sweet "incense beaten small," and entered "the Holy Place" from "the Court of the Tabernacle," and emptied his censer upon the Golden Altar, and "put the incense upon the fire before the Lord, that "the cloud of incense may cover the mercy-seat."[2] It was death for the priests to make the sacred incense for themselves, even "to smell thereto,"[3] or for any one but the priests, "the seed of Aaron," to offer it,[4] or to burn "strange incense" upon the Golden Altar.[5] Neither was it lawful to offer burnt sacrifice or meat-offering, or to pour drink-offering thereon;[6] but the blood of the sin-offering of atonements was once in every year sprinkled upon the horns thereof.[7] "Pure frankincense" was put on each of the two rows in which the twelve baked cakes of fine flour were set on the Table of Shewbread;[8] it was also put upon the meat-offering,[9] and was expressly prohibited to be put upon the sin-offering of fine flour, "for it is a sin-offering,"[10] or on "the offering of jealousy" of barley meal, "for it is an offering . . . . bringing iniquity to remembrance."[11] And in the service of the sanctuary, certain Levites were appointed to oversee "the frankincense and the spices,"[12] which were laid in "a great chamber" "of the House of God."[13]

Frankincense is often named in the poetical books of the Old Testament:—"Who is this that "cometh out of the wilderness, like pillars of smoke, perfumed with myrrh and frankincense, with "all powders of the merchant?"[14]—an image which those only can fully appreciate who have themselves been

> "Where through the sand of morning land
> The camel bears the spice."

"A garden inclosed is my sister, my spouse; a spring shut up, a fountain sealed. Thy plants are an orchard of pomegranates, with pleasant fruits; camphire [henna], with spikenard, spikenard and saffron; calamus [*Roosa* of India] and cinnamon, with all trees of frankincense; myrrh and aloes [aloeswood], with all the chief spices. A fountain of gardens, a well of living waters, and streams from Lebanon."[15]

"I have not caused thee to serve with an offering, nor wearied thee with incense [frankincense]. "Thou hast bought me no sweet cane [*Roosa*] with money."[16] "The multitude of camels shall "cover thee, the dromedaries of Midian and Ephah; all they from Sheba shall come; they shall "bring gold and incense" [frankincense].[17]

---

[1] Exod. xxx. 34-36.
[2] Lev. xvi. 12, 13.
[3] Exod. xxx. 37, 38.
[4] Numb. iii. 10; xvi.; 2 Chron. xxvi. 16-24.
[5] Exod. xxx. 9; Lev. x. 1-7; Numb. iii. 4, xxvi. 61.
[6] Exod. xxx. 9.
[7] Exod. xxx. 10; Lev. iv. 7.
[8] Lev. xxiv. 7.
[9] Lev. ii. 1, 2, 15, 16; vi. 15.
[10] Lev. v. 11.
[11] Numb. v. 15.
[12] 1 Chron. ix. 29.
[13] Nehemiah, xiii. 5-9.
[14] Song of Songs, iii. 6.
[15] Song of Songs, iv. 12-15.
[16] Isaiah, xliii. 23, 24.
[17] Isaiah, lx. 6.

"See a long race thy spacious courts adorn;
See future sons and daughters yet unborn,
In crowding ranks on every side arise,
Demanding life, impatient for the skies!
See barbarous nations at thy gates attend,
Walk in thy light, and in thy temple bend!
See thy bright altars throng'd with prostrate kings,
And heaped with products of Sabæan springs!
For thee Idume's spicy forests blow,
And seeds of gold in Ophir's mountains glow!"

"He that offereth an oblation is as if he offered swine's blood; he that burneth incense [frankincense], as if he blessed an idol."[1]

"To what purpose cometh there to me incense [frankincense] from Sheba, and the sweet cane " [*Roosa*] from a far country."[2] "And they shall come from the cities . . . . and from the plain, " and from the mountains, and from the south [Arabia[3]], bringing burnt-offerings, and sacrifices, " and meat-offerings and incense [frankincense], and bringing sacrifices of praise, unto the house " of the Lord."[4]

These passages emphatically derive frankincense from Sheba.

The Hebrew words for incense are quite different from the Hebrew word for frankincense (*Lebonah*); but in most of the passages where they are found frankincense may be understood as necessarily a constituent of the sacred incense of the Jewish ritual. Kitto says that these words for incense (*miktar*, *kitter*, and *keturoth*) all signify to raise an odour by burning, and are applied not only to the offering of incense, but of sacrifices, and he points out that the word which denotes the incense of sweet spices in Exod. xxx., 1, describes an incense of fat in Psalm lxvi., 15: "I will " offer unto thee burnt sacrifices of fatlings, with the incense of rams; I will offer bullocks with " goats." But the word may be used here to give poetic expression to the passage, as the thing itself was to give a sweet savour to the burnt sacrifices. In 1 Chron. vi., 49, the word for incense is also used, according to Calmet, for the fat of victims offered on the altar of the burnt offerings, but here, it may be supposed, without poetical licence. Gradually, as the spiritual discernment of the Jews developed, the offering of incense itself sublimed into prayer, as indicated in the figurative language of David, in Psalm cxli., 2, "Let my prayer be set forth before thee as incense, and the " lifting up of my hands as the evening sacrifice." And the same figure is used in the magnificent imagery of the Apocalypse: "And the four-and-twenty elders . . . . having every one of them " harps, and golden vials full of odours, which are the prayers of the saints."[5] "And another " angel came and stood at the altar, having a golden censer; and there was given unto him much " incense, that he should offer it with the prayers of all saints upon the golden altar which was " before the throne. And the smoke of the incense, with the prayers of the saints, ascended up " before God out of the angel's hand."[6]

The Jews buried their dead, the burning of the bodies of Saul and his sons[7] having been quite exceptional; but still they largely used unguents and spices, probably including frankincense, in their sepulchral rites. The dead body was anointed with spikenard,[8] and "they wound it in linen " clothes with the spices, as the manner of the Jews is to bury."[9] And of the burial of Asa it is written, "And they buried him in his own sepulchre, which he had made for himself in the city of " David, and laid him in the bed which was filled with sweet odours and divers kinds prepared by " the apothecary's art; and they made a very great burning for him."[10]

The account in the Iliad, of the burning of the dead body of Patroclus, would, by substituting the presentation of cows to Brahmins for the sacrifice of horses at the pyre, and milk for wine to quench its embers, serve to describe the funeral rites of a wealthy high-caste Hindoo of Bombay, in which frankincense, with all manner of balsamic substances, and fragrant woods and oils, are largely used; and we know that in the burning of their dead, and in their sacrifices, the Greeks, like the Romans, used frankincense extravagantly; but yet frankincense is not anywhere named in the Iliad or Odyssey, and it is said that it was unknown to the Greeks at the time when they were first sung.[11] But from about B.C. 400, and probably as the result of the opening of the Egyptian ports by Psammetichus, B.C. 670, and the destruction of Tyre, B.C. 573—two most important dates in the history of eastern products—frankincense is commonly and unceasingly mentioned by classical writers.

HERODOTUS (born B.C. 484) mentions it frequently;[12] and what he says about it is of course particularly valuable. He says, "Arabia . . . . is the only country which produces frankincense, " myrrh, cassia, cinnamon, ladanum. The Arabians do not get any of these, except myrrh, without " trouble. The frankincense they procure by means of the gum storax, which the Greeks obtain " from the Phœnicians; this they burn and thereby obtain the spice. For the trees which bear " the frankincense are guarded by winged serpents, small in size and of varied colours, whereof " vast numbers hang about every tree. They are of the same kind as the serpents which invade " Egypt, and there is nothing but the smoke of the styrax which will drive them from the trees."[13] " As one proceeds beyond Heliopolis up the country, Egypt becomes narrow, the Arabian range of " hills, which have a direction north to south, shutting it up on one side, and the Libyan range upon " the other. The former ridge runs on without a break and stretches away to the sea called " Erythræan; it contains the quarries whence the stone was cut for the pyramids of Memphis; and " this is the point where it ceases its first direction, and bends away in the manner above indicated. " In its greatest length from east to west it is, as I have been informed, a distance of two months' " journey; towards its extreme east its skirts produce frankincense."[14] "The Arabs brought every " year a thousand talents of frankincense, in tribute to Darius."[15] "It is also on the great altar " [of gold] that the Chaldæans burn the frankincense, which is offered to the amount of one thou- " sand talents weight, every year at the festival of the god."[16] [Bel]. In describing "the mode of " embalming amongst the Egyptians, according to the most perfect practice," he says "they fill

---

[1] Isaiah, lxvi. 3.
[2] Jeremiah, vi. 20.
[3] Matt. xii. 42; compare Gen. xxv. 6, and Matt. ii. 1.
[4] Jeremiah, xvii. 26.
[5] Rev. v. 8.
[6] Rev. viii. 3, 4.
[7] 1 Sam. xxxi. 12, 13
[8] John, xii. 7.
[9] John, xix. 39, 40.
[10] 2 Chron. xvi. 14; Pet. Cunæus de Respub. Hebræorum. Leyden, 1631.
[11] Crusius, Homeric Lexicon, Translation Smith, ed. Arnold.
[12] Schweighæuser, Lexicon Herodoteum.
[13] Book iii. Thalia, 107, Rawlinson's Translation.
[14] Book ii. Euterpe, 8, Rawlinson's Translation.
[15] Book iii. Thalia, 97, Rawlinson's Translation.
[16] Book i. Clio, 183, Rawlinson's Translation.

" the cavity [of the abdomen] with the purest bruised myrrh, with cassia, and every sort of " spicery, *except frankincense*."[1] And of the Scythians, " Their women make a mixture of cypress, " cedar, and frankincense wood, which they pound into a paste upon a rough piece of stone, adding " a little water to it. With this substance, which is of a thick consistency, they paste their faces " all over, and indeed their whole bodies. A sweet odour is thereby imparted to them ; and when " they take off the plaster on the day following, their skin is clean and glossy."[2]

THEOPHRASTUS, who lived only a century later (B.C. 394–287), gives the fullest and most accurate account of frankincense of all ancient writers. I give a translation of what he says about it, in full, and especially as it has never before, I believe, been done into English (History of Plants, book ix. ch. 4)[3].

" Concerning frankincense and myrrh and balsam, and whatever else is like these, it has [already] been said, that they are produced by incision, and spontaneously. And we must [now] endeavour to tell what is the nature of the trees, and if they have anything peculiar as to their origin or collection, or other matters; and, in like manner, concerning the other sweet-smelling trees; for almost the whole of them grow in places towards the south and east. The frankincense-tree and myrrh and cassia and cinnamon grow in the Chersonese of the Arabians, about Saba and Adramyta, and Citibæna and Mali. But the trees of frankincense and myrrh grow, some of them on the mountain, and others in private plantations, at the foot of the mountain; on which account, some of them are cultivated, and others are not: and they say that the mountain is lofty and thickly wooded, and covered with snow, and that rivers also flow down from it into the plains, and that the frankincense-tree is not large, being five cubits high, and covered with boughs, and that it has a leaf like that of the pear-tree, only much smaller, and is of a grassy colour, very like rue, and has altogether a smooth bark like the laurel; but that the myrrh-tree is still smaller in size and more shrub-like, and that it has a hard trunk, and is twisted towards the ground, and is thicker than a man's leg, and has a smooth bark, like purslane. But others, who say they have seen them, nearly all agree concerning their size, namely, that neither of the trees is large, the myrrh-tree being the smaller and lower [of the two]. And they state that the frankincense bears a resemblance to a laurel, and that it has a smooth bark, but that the myrrh is prickly and not smooth, and that it has a leaf like the elm, only crisp, and prickly at the top, like the ilex-tree. And these said that in a voyage which they were making from the Bay of Heroes, they disembarked to search for water on the mountain, and thus saw the trees and the mode of collecting [the frankincense]. And that the trunks and boughs of both were incised; but that the former appeared to have been cut, as it were, by an axe, and the latter to have had more gentle incisions ; and that the drops partly fell down and partly remained on the tree. And that in some places mats woven of palm-leaves were placed underneath, while in others the ground underneath was hardened and kept clean ; and that the frankincense on the mats was pure and transparent, but that on the ground less so ; and that they scraped off what adhered to the trees with knives, so that the bark stuck to some of them. And they said that the whole mountain was divided among the Sabæans ; for that they were the lords [of the place], and that they were just towards one another, on which account no one kept any guard [over his own property]; and that having themselves taken thence an abundance of frankincense and myrrh, which they placed in their ships, none of the inhabitants being present, they had sailed away. And these both told this, and said they heard that the myrrh and frankincense is gathered together on every side to the temple of the Sun; and that this belongs to the Sabæans, being by far the most sacred thing in the country, and that certain armed Arabs have the custody of it; and that when they bring it, each, heaping up his own frankincense, and the myrrh in a similar way, leaves them with the keepers, and places upon the heap a little tablet, stating the number of the measures, and the price at which each measure is to be sold ; and that when the merchants come they inspect these tablets, and having measured any heap that pleases them they put the price of it in the place from which it is taken ; and that the priest then coming takes a third part of the price for the god, and there leaves the remainder, which is kept safe for the owners until they come and take it. But certain others say that the frankincense-tree is like the lentisck, and its fruit to the berries of the same, and that the leaf of it is reddish; and that the frankincense from the young trees is whiter and less fragrant, while that from the older trees is yellowish and more fragrant; and that the myrrh-tree is like the terebinth, but rougher and more thorny, and the leaf a little rounder and, if chewed, resembling the terebinth in taste; and that of these, also, the older are the more fragrant. And that both grow in the same place, and that the ground [there] is argillaceous and flaky, and that springs of water are rare. These things, however, are contradictory [to the statement] that it snows and is wet [in that locality], and that rivers issue from it. And others also say that the tree is like the terebinth, and others that it is the terebint hitself; for that specimens of the wood were brought to Antigonus by the Arabs who conveyed the frankincense, and that they differed in nothing from the terebinth. These, however, showed still greater ignorance; for they thought that both the frankincense and the myrrh grew on the same tree. On which account, the report brought by those that sailed from the City of Heroes is more credible ; since the frankincense-tree that grew above Sardis, in a certain temple, has a leaf like the laurel, if from this we may form a conclusion, and the frankincense produced from it, whether from the trunk or branches, is like the other frankincense in appearance, and in smell when it is burnt. And this tree alone grew without [any culture]. And some say that the frankincense grows more abundantly in Arabia, but more beautiful in the neighbouring islands,

---

[1] Book ii. Euterpe, 86, Rawlinson's Translation.

[2] Book iv. Melpomene, 75, Rawlinson's Translation.

[3] Θεόφραστος. Theophrasti Eresii opera quæ supersunt omnia. Græca recensuit, Latine interpretatus est, indices rerum et verborum absolutissimos adjecit Fridericus Wimmer, Doct. Philos. Parisiis, Didot, 1866, 8vo., pp. 143–145.

over which the Arabs have sway; for there they make figures upon the trees of whatever they like; which is not incredible, as they admit of any incision that persons may wish to make in them. Some of the grains also are very large, in bulk as much as a handful, and in weight more than the third part of a mina. All frankincense is brought to market in a rough state, similar in appearance to the bark of a tree; but of myrrh there are two kinds, the one in drops, and the other in moulds. The quality is judged of by the taste; and from this they choose what is of uniform colour. Concerning frankincense and myrrh, this is nearly as much as we have heard up to the present time."

A fragment of AGATHARCIDES (B.C. 200) is said, for I have not seen the passage myself, to enumerate balm, cassia, frankincense, myrrh, and cinnamon as the production of Saba.

DIODORUS (about B.C. 50) writes[1]:—"Next to these inhabit those Arabians called Carbi, and next to them the Sabæans, the most populous of any of the Arabians; for they possess Arabia the Happy, exceeding rich in all those things which we esteem most precious; and for breeding of cattle of all sorts, the most fertile country in the world; for the whole country is naturally perfumed all over, almost everything growing there sending forth continually most excellent odours. On the sea coasts grow balsam, cassia, and another herb of a strange and peculiar property, which while it is fresh is delightful and strengthening to the eyes, but kept awhile, presently loses its virtue. Higher in the heart of the country, are shady woods and forests, graced and beautified with stately trees of frankincense and myrrh, palm-trees, calamus, and cinnamon, and such like odoriferous plants; for none can enumerate the several natures and properties of so great a multitude, or the excellency of those sweet odours that breathe out of every one of them. For their fragrancy is such, that it even ravishes the senses with delight, as a thing divine and unutterable; it entertains them that sail along by the coast at a great distance with its pleasures and delights. For in spring-time the winds from off the land waft the air, perfumed with the sweet odours of myrrh and other odoriferous plants, to those parts of the sea that are next to them. And these spices have nothing of a faint and languishing smell, as those that come to our hands, but a strong and vigorous odour, that strongly pierces all their senses to the utmost of their capacity; for the wafts of air dispersing the perfumes of these odoriferous plants, abundance of pleasant, healthful, and strange variety of scents (proceeding from the richest spices) are conveyed to them that sail near unto the coast. For this sweet smell comes not from fruit bruised in a mortar (whose strength is in a great measure decayed) or from spices made up in divers sorts of vessels for transportation; but from the ripeness of the fruit as it grows, and from the pure and divine nature of the plant itself. So that they that have the advantage of these sweet odours, seem as if they were entertained with that feigned meat of the gods called ambrosia, since those excellent perfumes cannot have a name ascribed them transcending their worth and dignity.

"Yet fortune has not imparted to men an entire and unmixed felicity in these things; but has joined some inconveniences with these advantages, to correct them who (through a constant confluence of earthly blessings) have usually despised and slighted the gods. For these fragrant forests abound with red serpents of a span long, whose bite is deadly and incurable. They strike a man with a violent assault, leaping up in his very face, leaving him besmeared with his own blood.

"The capital city of this nation is called Saba, and stands upon a high hill; they are governed by kings, who inherit the crown by descent; yet the honour allowed them by their subjects is such, as that they are in some respects, notwithstanding, in a sort of bondage and slavery; for, though they seem to be privileged in this, that they have a sovereign and absolute power in making laws, and are not to give an account of any of their actions to their subjects; yet they are as unhappy in this, that they are never to stir out of their palace; for if they do, the people stone them to death, for so they are commanded by an ancient oracle.

"This nation not only excels all the neighbouring barbarians in wealth, but all other people whatsoever for plenty of every thing that is accounted precious; for in their traffic for a thing of a very small weight, they receive a greater sum of money than any other merchants that sell goods for silver.

"Being, therefore, that they never were conquered, by reason of the largeness of their country, they flow, as it were, in streams of gold and silver, especially at Saba, the seat royal of their kings; their vessels and all their cups are of gold and silver, and likewise their beds, chairs, and stools have their feet of silver; and all other their household stuff is so sumptuous and magnificent that it is incredible. The porticoes of their houses and temples are some of them overlaid with gold; and silver statues are placed upon some of the chapiters of the temples. The doors and roofs of their inner rooms are adorned with many golden bowls, set with precious stones, and other things of great value; for they have enjoyed a constant and uninterrupted peace for many ages and generations, being very far remote from those whose covetousness prompts them to advance their gain by the riches of others.

"The sea there is of a very white colour, so that a man may as justly wonder at the strangeness of the thing, as be inquisitive what should be the cause.

"Near hereunto are the Fortunate Islands, full of walled towns, where all the sheep and cattle are exceeding white, and none of the females bear any horns. The merchants from all parts resort to these islands, especially from Pontana, built by Alexander the Great, upon the banks of the river Indus, that there might be a commodious port-town for shipping upon these coasts; but of this country and its inhabitants, we have said enough."

---

[1] Book v. chap. iii. Booth's Translation, Lond. 1814, vol. i. pp. 186–189.

And [1]:—"Having gone through the western and northern countries and the islands of the ocean, we shall now describe the southern islands, lying in the Arabian ocean, on the east part of Arabia, next to Gedrosia. This part of Arabia is a country full of villages and considerable towns, some of which are situated upon high hills, others upon rising grounds, or something higher than champaign fields. Their greatest cities have stately royal palaces, and are very wealthy and populous; the country abounds with all sorts of cattle, and is of a very fruitful soil, affording plenty of rich pastures for the flocks and herds; many rivers run through it, watering the fields, to the great increase of the fruits of the earth. And, therefore, this part of Arabia, which excels the rest in richness of soil, is justly called Arabia the Happy.

"Over against the utmost point of this country, near the ocean, lie many islands; but there are but three that are worth remark: the first is called the Holy Island, wherein it is unlawful to bury the dead; but not far from this, about seven furlongs distant, there is another, wherein they bury. The sacred isle produces frankincense, and in that abundance as suffices for the service and worship of the gods all the world over; it has likewise plenty of myrrh, with other odoriferous spices of several sorts, which breathe out a most fragrant smell. The nature of frankincense, and the manner of getting it, is thus: the tree is very small, like to the white Egyptian thorn, and bears a leaf like to the willow; it puts forth a flower of a golden colour; from the bark of this tree, by incision made, distils the frankincense in drops, like tears.

"The myrrh-tree is like to the mastic-tree, but bears a more slender leaf and grows thicker on the branches. The myrrh flows forth by digging up the earth round about the roots. Those that grow in a rich soil, bear twice a year—that is, in the spring and summer; that in the spring-time is of a deep red colour, caused by the dew; the other, nearer winter, is white.

"There they got likewise the fruit of the paliurus tree, wholesome both in meat and drink, and good against a dysentery. The land is divided amongst the inhabitants, of which the best part is allotted to the king, who has likewise the tenths of the fruits.

"They say the breadth of this island is about two hundred furlongs, inhabited by them they call Panchæans, who transport the myrrh and frankincense into foreign parts, and sell it to the Arabian merchants, from whom others buy these and such like merchandise, and convey them to Phœnicia, Cœlesyria, and Egypt; and from those places they are carried by the merchants over all parts of the world.

"Besides these, there is another large island, about thirty furlongs distance from this last mentioned, lying to the east, many furlongs in length. For they say, from a promontory thereof running out towards the east, may be seen India, like a cloud in the air, the distance is so great.

"There are many things observable in Panchæa that deserve to be taken notice of. The natural inhabitants are those they call Panchæi; the strangers that dwell among them are people of the western parts, together with Indians, Cretans, and Scythians.

"In this island there is a famous city, called Panara, not inferior to any for wealth and grandeur. The citizens are called the suppliants of Jupiter Triphylius, and are the only people of Panchæa that are governed by a democracy without a monarch. They choose every year the presidents or governors, that have all matters under their cognizance; but what concerns life and death, and the most weighty matters, they refer to the college of their priests. The temple of Jupiter Triphylius is about sixty furlongs distant from the city, in a champaign plain. It is in great veneration, because of its antiquity and the stateliness of the structure, and the fertility of the soil.

"The fields round about the temple are planted with all sorts of trees, not only for fruit, but for pleasure and delight; for they abound with tall cypresses, plane trees, laurels, and myrtles, the place abounding with fountains of running water; for near the temple there is such a mighty spring of sweet water rushes out of the earth, as that it becomes a navigable river; thence it divides itself into several currents and streams, and waters all the fields thereabouts, and produces thick groves of tall and shady trees; amongst which, in summer, abundance of people spend their time, and a multitude of all sorts of birds build their nests, which create great delight both by affecting the eye with the variety of their colours, and taking the ear with the sweetness of their notes. Here are many gardens, sweet and pleasant meadows, decked with all sorts of herbs and flowers, and so glorious is the prospect, that it seems to be a paradise worthy of the gods themselves.

"There are here likewise large and fruitful palms, and abundance of walnut-trees, which plentifully furnish the inhabitants with pleasant nuts. Besides all these, there are a multitude of vines of all sorts, spiring up on high, and so curiously interwoven one amongst another, that they are exceeding pleasant to the view, and greatly advance the delights of the place.

"The temple was built of white marble, most artificially jointed and cemented, two hundred yards in length, and as many in breadth, supported by great and thick pillars, curiously adorned with carved work. In this temple are placed huge statues of the gods, of admirable workmanship, and amazing largeness. Round the temple are built apartments for the priests that attend the service of the gods, by whom everything in that sacred place is performed. All along from the temple is an even course of ground, four furlongs in length, and a hundred yards in breadth; on either side of which are erected vast brazen statues, with four square pedestals; at the end of the course breaks forth the river, from the fountains before mentioned, from whence flows most clear and sweet water, the drinking of which conduces much to the health of the body. This river is called the water of the sun.

[1] Book v. chap. iii. Booth's Translation, Lond. 1814, vol. i. pp. 324–329.

"The whole fountain is lined on both sides, and flagged at the bottom with stone, at a vast expense, and runs out on both sides for the space of four furlongs. It is not lawful for any body but the priests to approach to the brink of the fountain. All the land about, for two hundred furlongs round, is consecrated to the gods, and the revenues bestowed in maintaining the public sacrifices and service of the gods; beyond these consecrated lands is a high mountain, dedicated likewise to the gods, which they call the throne of Cœlus and Triphylius Olympus; for they report that Uranus, when he governed the whole world, pleasantly diverted himself in this place, and from the top of the mount observed the motions of the heavens and stars, and that he was called Triphylius Olympus, because the inhabitants were composed of three several nations, Panchæans, Oceanites, and Doians, who were afterwards expelled by Ammon; for it is said that he not only rooted out this nation, but utterly destroyed all their cities, and laid Doia and Asterusia even with the ground. The priests every year solemnize a sacred festival in this mountain, with great devotion.

"Behind this mount, in other parts of Panchæa, they say there are abundance of wild beasts of all kinds, as elephants, lions, leopards, deer, and many other wonderful creatures, both for strength and proportion. In this island there are three chief cities, Hyracia, Dalis, and Oceanis. The whole country is very fertile, and especially in the production of all sorts of wine in great plenty.

"The men are warlike, and use chariots in battle, after the ancient manner. The whole nation is divided into three parts: the first class is of the priests, with whom are joined the artificers; the other tribe consists of the husbandmen; and the third are the militia and the shepherds.

"The priests govern all, and are the sole arbitrators in every matter; for they give judgment in all controversies, and have the power and authority in all public transactions of state. The husbandmen till the land; but the fruit is brought into the common treasury; and who is judged the most skilful in husbandry, receives the largest share of the fruits for a reward in the first place; and so the second, and the rest in order to the tenth, as every one merits, less or more, receives his reward by the judgment of the priests. In the same manner the shepherds and herdsmen carefully bring into the public stock the victims and other things, both by number and weight, as the nature of things are; for it is not lawful for any to appropriate any thing to themselves particularly, except a house and garden. For all the young breed of cattle and other things, and all the revenues, are received by the priests, and they justly distribute to every one as their necessity does require; only the priests have a double proportion.

"They wear soft and fine garments; for the sheep's wool is much finer here than anywhere else. Both men and women likewise deck themselves with golden ornaments; for they wear necklaces of gold, and bracelets about their arms, and, like the Persians, have earrings hanging in their ears. Their shoes are such as others wear, but richly beautified with divers sorts of colours.

"Their soldiers, for ordinary pay, defend the country, fortifying themselves within camps and bulwarks; for there is a part of the island infested with most daring thieves and robbers, who often lurch and surprise the husbandmen.

"To conclude, these priests, for delicacy, state, and purity of life, far excel all the rest of the inhabitants; their robes are of white linen, and sometimes of pure soft wool. They wear likewise mitres embroidered with gold. Their shoes are sandals curiously wrought with exquisite workmanship; and in their ears hang golden earrings like to the women's.

"They attend chiefly upon the service of the gods, singing melodious songs in their praises, setting forth their glorious acts and benefits bestowed upon men. The priests say they came originally from Crete, and were brought over into Panchæa by Jupiter when he was upon earth and governed all the world; and allege their language for a confirmation of this assertion, inasmuch as they retain many words of the Cretan speech among them: and further say that they derive from their ancestors that civility and kindness wherewith they entertain the Cretans, the fame and report of their ancient consanguinity descending continually in a perpetual succession to their posterity; they show likewise a record written, as they say, by Jupiter's own hand, at the time when he was on earth and laid the foundation of the temple.

"There are in this island likewise mines of gold, silver, brass, and iron; but it is not lawful for any to export them. Nay, it is not lawful for any of the priests to go out of the verge of the consecrated ground; and if any do, it is lawful for any man that finds them to kill them. They have under their charge innumerable vast vessels, and other consecrated things, both of gold and silver, which have been laid up there in honour of the gods for many ages. The gates of the temple are of admirable workmanship, beautified with gold, silver, ivory, and thyine wood.

"The bed of the god is six cubits long, and four broad, of massy gold, most curiously wrought in every part; near adjoining, stands the table, as large, and of the like materials and workmanship with the other in every respect.

"In the middle of the bed is placed a great golden pillar, whereon are letters inscribed, called by the Egyptians sacred writing, expressing the famous actions of Uranus, Jupiter, Diana, and Apollo, written, they say, by Mercury himself. But this may suffice concerning the islands lying in the ocean over against Arabia."

STRABO (B.C. 54—A.D. 24) says,[1] "Frankincense is produced in Catabania and in the parts of "Africa opposite. Here the frankincense-tree grows along the banks of the Isis and Nilus. The "country of the Sabæi produces not only frankincense, but balsam, sweet-smelling palms, calamus, "and larimnum, a most fragrant perfume. By the trade in these aromatics both the Sabæans and

---

[1] Book xvi. chap. iv. Falconer and Hamilton's Translation, Bohn's Classical Lib.

" the Gerrhæi have become the richest of all the tribes, and possess a great quantity of wrought
" articles in gold and silver, as couches, tripods, basins, drinking-vessels; to which we must add
" the costly magnificence of their houses; for the doors, walls, and roofs are variegated with inlaid
" ivory, gold, silver, and precious stones." The expedition sent by Augustus, under the command of Ælius Gallus, against the Arabians (B.C. 24–A.D. 1) (at the same time that Petronius was sent into Æthiopia, and reduced Candace, queen of Meroe), was to explore the nature of these places and their inhabitants. It was very unfortunate. The expedition was fitted out as if for a great war; but the cities yielded one after another without a struggle, and although the army suffered exceedingly from long marches over bad roads, and from want of food, and boils, only seven men fell in the expedition, which "contributed little in extending our knowledge of the country." "The
" aromatic country is divided into four parts. Of aromatics, the frankincense and myrrh are said
" to be the produce of trees, but cassia of bushes. Yet some writers say that the greater part of it
" [the cassia] is brought from India, and that the best frankincense is that of Persia." "The Naba-
" tæans worship the sun, and construct the altar on the top of the house, pouring out libations, and
" burning frankincense on it every day."

PLINY (A.D. 23-79) says,[1] "There is no region in the whole world that bringeth forth frankin-
" cense but Arabia; and yet it is not to be found in all parts thereof, but in that quarter only of
" the Atramites. Now these Atramites inhabit the very heart of Arabia, and are a country of the
" Sabæi. The capitall city of the whole kingdom is called Sabota, seated upon a high mountain;
" from whence into Saba, the only country that yields such plenty of the said incense, it is about
" eight daies journey. As for Saba (which in the Greek tongue signifieth, a secret mysterie) it
" regards the Sunne rising in summer, or the north-east, enclosed on every side with rocks inacces-
" sible; and on the right hand it is defended with high cliffes and crags that beare into the sea.
" The soile of this territorie, by report, is reddish, and inclining to white. The forrests that carry
" these incense-trees ly in length twenty schœnes, and beare in bredth half as much. Now that which
" we cal schœnus, according to the calculation of Eratosthenes, contains forty stadia, that is to say,
" five miles; howsoever some have allowed but thirty-two stadia to every schœnus. The quarter
" wherein these trees grow is full of high hills; howbeit, go down into the plains and valley
" beneath, you shall have plenty of the same trees, which come up of their own accord, and were
" never planted. The earth is fat, and standeth much upon a strong clay, as all writers do agree.
" Few springs are there to be found, and those that be are full of nitre. There is another tract by
" it selfe confronting this country, wherein the Minæans do inhabit; and through them there is a
" narrow passage, whereby the frankincense is transported into other parts. These were their first
" neighbours that did traffique with them for their incense, and found a vent for it; and even so
" they doe still at this day, whereupon the frankincense itselfe is called of their name, Minæum.
" Setting these people of the Sabeans aside, there be no Arabians that see an incense-tree from one
" end of the yere to another; neither are all these permitted to have a sight of those trees. For
" the common voice is, that there be not above 3,000 families which can claime and challenge by
" right of succession that priviledge to gather incense. And therefore all the race of them is called
" sacred and holy; for looke when they go about, either cutting and flitting the trees, or gathering
" the incense, they must not that day come neere a woman to know her carnally; nay, they must
" not be at any funerals, nor approach a dead corps, for being polluted. By which religion and
" ceremonious observation the price is raised and the incense is the dearer. Some say, these people
" have equall liberty in common to go into these woods for their commodities when they will; but
" others affirme that they be divided into companies, and take their turns by yeares. As concern-
" ing the tree, I could never know yet the perfect description of it. We have waged warres in
" Arabia, and our Roman armie have entered a great way into that country. C. Cæsar, the adopted
" son of Augustus, wan great honour and glory from thence; and yet verily, to my knowledge,
" there was never any Latine author that hath put down in writing the form and fashion of the
" tree that carrieth incense. As for the Greeke writers, their bookes doe vary and differ in that
" point. Some give out, that it hath leaves like to a peare tree, only they be somewhat lesse, and
" when they come forth they be of a grasse green colour. Others say that they resemble the
" lentiske tree, and are somewhat reddish. There be again who write, that it is the very terebints
" and none else, that giveth the frankincense: of which opinion king Antigonus was, who had one
" of these shrubs brought unto him. King Juba in those books which he wrot and sent to C. Cæsar,
" son to the Emperor Augustus (who was inflamed with an ardent desire to make a voiage into
" Arabia, for the great fame that went thereof), saith, That the tree which beares frankincense hath
" a trunke or body writhen about, and putteth forth boughes and branches, like for all the world to
" the maple of Pontus. Item, that it yeeldeth a juice or liquor as doth the almond tree; and such
" are seene commonly in Carmania; as also those in Egypt, which were planted by the careful
" industrie of the Ptolomees, kings there. However it be, this is received for certaine, that it hath
" the very barke of a bay tree; some also have said that the leaves be as like. And verily such
" kind of trees were they which were seen at Sardis; for the kings of Asia likewise were at the
" cost and labour to transplant them, and desirous to have them grow in Lydia. The embassadours
" who, in my time, came out of Arabia to Rome, have made all that was delivered as touching
" these trees, more doubtfull and uncertaine than before. A strange matter, and wonderfull indeed,
" considering that twigges and branches of the incense tree have passed betweene; by the view of
" which impes, we may judge what the mother is: namely, even and round in the body, without
" knot or knar, and from thence she putteth out shoots.

[1] Book xii. chap. 14. Holland's Translation, Lond. 1635, fol.

"They used in old time to gather the incense but once a yere, as having little vent, and small returne, and lesse occasion to sell than now adaies; but now, since every man calleth for it, they feeling the sweetnesse of the gaine, make a double vintage (as it were) of it in one yere. The first, and indeed the kindly season, falls about the hottest daies of the summer, at what time as the Dog daies begin; for then they cut the tree where they see the bark to be fullest of liquor, and whereas they perceive it to be thinnest and strut out most. They make a gash or slit only to give more libertie; but nothing do they pare or cut cleane away. The wound or incision is no sooner made, but out there gusheth a fat fome or froth; this soon congeales and growes to be hard; and where the place will give them leave, they receive it in a quilt or mat, made of date-tree twigs, plaited and wound one within another wicker-wise. For elsewhere, the floore all about is paved smooth, and rammed downe hard. The former way is the better to gather the purer and clearer frankincense; but that which falleth upon the bare ground prooves the weightier. That which remaines behind, and stickes to the tree, is parted and scraped off with knives, or such like yron tooles, and therefore no marvell if it be full of the shavings of the bark. The whole wood or forrest is divided into certaine portions; and every man knowes his owne part; nay, there is not one of them that will offer wrong unto another, and encroch upon his neighbors. They need not to set any keepers to look unto those trees that be cut, for no man will rob from his fellow if he might, so just and true they be in Arabia. But, believe me, at Alexandria, where frankincense is tried, refined, and made for sale, men cannot look surely ynough to their shops and workhouses, but they will be robbed. The workemen that is emploied about it is all naked, save that he hath a paire of trouses or breeches to cover his shame, and those are sowed up and sealed too, for fear of thrusting any into them. Hood-winked he is sure enough for seeing the way to and fro, and hath a thicke coife or maske about his head, for doubt that he should bestow any in mouth or eares. And when these workmen be let forth againe, they be stripped starke naked, as ever they were borne, and sent away. Whereby we may see, that the rigor of justice cannot strike so great feare into our theeves here, and make us so secure to keepe our owne, as among the Sabæns the bare reverence and religion of those woods. But to returne again to our former cuts. That incense which was let out in summer, they leave there under the tree until the autumne, and then they come and gather it. And this is most pure, cleane and white.

"A second vintage and gathering there is in the spring; against which time they cut the barks before in the winter, and suffer it to run out until the spring. This comes forth red, and is nothing comparable to the former. The better is called Carpheotum; the worse, Dathiatum. Moreover, some say, that the gum which issueth out of the young trees is the whiter; but that which comes from the old, is more odoriferous. There be others also of opinion, that the better incense is in the islands. But King Juba doth avouch constantly, that there is none at all in the islands. That which is round, like unto a drop, and so hangeth, we call the male incense, whereas in other things lightly we name the male, but where there is a female. But folk have a religious ceremonie in it, not to use so much as the tearme of the other sexe in giving denomination to frankincense. Howbeit, some say that it was called the male for a resemblance that it hath to cullions or stones. In very truth, that is held for the cheife and best simply, which is fashioned like to the nipples or teats that give milk, standing thick one by another; to wit, when the former drop that distilled, hath another presently followeth after, and so consequently more unto them, and they all seem to hang together like bigs. I read that every one of these were wont to make a good handfull, namely, when men were not so hasty and eager to carry it away, but would give it time and leisure to drop softly. When it is gathered in this sort, the Greeks used to call it Stagonias and Atomus; but the lesser goblets they name Orobias. As for the small crums or fragments which fall off by shaking, we call manna (*i.e.* Thuris). And yet there be found at this day drops of incense that weigh the third part of a pound, that is to say, about thirty-nine Roman deniers[1]. It happened on a time that King Alexander the Great, being then but a very little child, made no spare of incense, but cast still upon the altar without all measure when he offered sacrifice. Whereupon Leonides, his tutor and schoolemaister, by way of a light reproofe, said unto him thus, Sir, you should in that maner burne incense, when you have once conquered those nations where there growes incense. Which rebuke and checke of his tooke so deep a print in Alexander's heart, and so well he carried it in memorie, that after he had indeed made conquest of Arabia, he sent unto the said Leonides, his tutor, a ship ful fraught and charged with incense; willing him not to spare, but liberally to bestow upon the gods when hee sacrificed. To returne againe to our historie. When the incense is gathered (as is beforesaid) conveighed it is to Sabota, upon cammel's backs, and at one gate (set open for that purpose) it is brought into the citie. For by law forbidden it is on pain of death to take any other way. Which done, the priests there of the god whom they call Sabis, take the disme or tenth part of the incense, by measure, and not by weight, and set it apart for that god. Neither is it lawful for any man to buy or sell, before that duty be paied; which serves afterwards to support certaine publick expenses of the citie. For all strangers and travellers within the compasse of certain daies journey, if they come to the citie are courteously received, and liberally entertained at the cost and charge of the said god Sabis. Carried forth of the country it cannot be, but through the Gebanites; and therefore there is a custome paid to their king. The head citie of that kingdom, Thomna, is from Gaza (the next port-towne in Judæa toward our coast) seven-and-twentie miles fourscore times told; and this way is divided into sixty-two daies journey by camels. Moreover, besides tyth aforesaid, there be measures, bestowed upon the priests to their owne use and others likewise to the king's secretaries and scribes; and not only these have a share, but also the keepers, sextons, and wardens of the temple,

[1] Or rather thirty-three and a scruple.

the squires of the bodie, the guard and pensioners, the king's officers, the porters, groomes, and other servitors, pill and poll, and every one hath a snatch. Moreover, all the way as they travell; in one place they pay for their water, in another for fodder and provender, or else for their lodging and stable-room, and every where for one thing or other they pay toll; so as the charge for every cammell from thence to the sea, upon our coast commeth to 688 deniers; and yet we are not come to an end of paiements, for our publicanes and customers also belonging to our empire must have a fleece for their parts. And therefore a pound of the best incense will cost sixteen deniers, of the second fifteen, and the third fourteen. With us it is mingled and sophisticated with parcels of a white kind of rosin, which is very like to it; but the fraud is soon found by the meanes above specified. The best incense is tried and knowne by these markes, viz.: if it be white, large, brittle, and easie to take a flame when it comes neare a coale of fire; last of all, if it will not abide the dent of the tooth, but flie in pieces and crumble, sooner than suffer the teeth to enter into it."

Pliny also says that the Sabæans cooked their food with the wood of the frankincense-tree; and that, "in the Trojan times, there were no ointments, neither was supplication made with frank- " incense "[2].

ARRIAN (about A.D. 90) writes[3], "Beyond Kana [Makalla] the bay called Sakhalites [Makalla Bay. " —G.B.] commences: it is of vast extent and of considerable depth, and forms the coast of the " frankincense country, a mountainous and impracticable tract, incommoded with fogs and a " dark atmosphere in all the parts where the trees grow that produce the frankincense. The trees " are neither large nor lofty; but the substance exudes from the bark, and becomes consistent, " like the gum that weeps from several of our trees in Egypt. The incense is collected by the " king's slaves, condemned to the service as a punishment; for the county is unhealthy in the " extreme, pestilential even to those who navigate on the coast, and certain death to the wretched " sufferers employed in the collection, who, if they escape death from the climate, are sure to perish " by want and neglect. The coast which forms the border of this country terminates at the " promontory called Suagros [Ras Fartak], the largest cape in the world, and projecting towards " the east. There is a garrison here for the protection of the country, and a harbour, on which are " the storehouses for the frankincense collected here from the whole coast." . . . "From Suagros the " adjoining bay of Omana [Bay of Al Kammar?—G.B.] takes a large sweep of considerable depth " into the main, six hundred stadia in extent, and towards the termination of this, the shore is " high, rocky and steep too, for five hundred stadia more. At this boundary is the established port " for the reception of the Sakhalitic incense, called Moskha [Morbat or Merbat?—G.B.]; it is a " place regularly frequented by the vessels from Kane [Makalla], whilst the fleets returning from " Barugaza [Broach] and Limurike [Canara and Malabar], if they happen to lose the season, are " obliged to lie here and exchange part of their cargo, such as wheat, oil (of sesamum), or cottons, " for frankincense, which is supplied by the king's agents and factors. Throughout the whole tract, " called Sakhalites, frankincense lies piled up in heaps, without watch or guard to secure it, as if " it were committed to some divine power for protection. For without the king's permission it is " impossible to convey it on ship-board, either openly or clandestinely; even if a single grain were " embarked it would be a marvellous chance if the vessel escaped seizure. From Moskha [Morbat " or Merbat?] the coast extends fifteen hundred stadia more to the district of Asikho [Hafek of El " Edresi]; and at the termination of this tract lie the seven islands of Zenobius [Curia Muria] in " succession."

Again, of the country opposite, between Mount Elephant [Ras Fiel] and Aromata [Gardafui], he writes, "In these parts, the best and greatest quantity of frankincense is procured."[4]

Arrian mentions ἀρώματα amongst the exports of Abaletes, and that the myrrh of the African coast was taken to the opposite coast of Arabia; and names frankincense amongst the exports of Malao (Zeyla?), Moundus, Mosullon, or Mossylon (Berbera), and of the ports of Barbaria. The frankincense of these latter ports he calls "Peratick," or "*foreign*." "But," observes Vincent[5], "it " cannot be admitted in that sense, as to the commodity itself, for it is noticed expressly as a native " produce of the place. Still it will lead us to solve a difficulty already noticed in regard to these " ports of Barbaria called *Ta pera*, which, by a slight correction of the text will signify the ports " beyond the straits (Bab-el-Mandeb). The articles obtained here would naturally be styled " *Peratick* (from Pera, *beyond*), and would be known by this title in the invoices, and the market " of Alexandria, in contradistinction to those obtained within the Straits of Bab-el-Mandeb. The " author is writing to Alexandrians, and is consequently specifying the precise parts where those " commodities were obtained which they knew by the name of Peratick." Arrian names frankincense also as an export of Tobai, near Aromata, and of Kane (Makalla) in Arabia. He names it also as an import of Barbarike, at the mouth of the Indus, amongst the exports of which he names costus, bdellium, spikenard, indigo, all exports of Kurrachee at the present day. We have already seen that he names frankincense as exported to Barygaza (Broach) and Limurike from Moskha (Merbat). But when speaking of the exports of Ozene (Ougein in Central India, where the *Boswellia thurifera* of Colebrooke abounds) to Barygaza (Broach)—onyx stones, murrhine vases, and cotton, &c. — he does not specify frankincense. Ozene transmitted also spikenard, costus, and bdellium from Upper India to Barygaza; and amongst the imports of Barygaza are named also lycium, storax, melilotus, myrrh, and other aromata and unguents.

---

1 Book xii. ch. 17, Celsius. 2 Book xiii. ch. 1, Celsius.
3 Vincent's Translation of the *Periplus and Voyage of Nearchus*, Oxford, 1809. 4 Ibid.
5 Vincent's "Commerce and Navigation of the Ancients in the Indian Ocean," Lond. 1807, vol. i. p. 140.

Ptolemy[1] (about A.D. 150) places the Libanotophoros, or thuriferous region, between Makalla and Muscat, placing the Smyrnophoros to its west, behind Makalla.

Dioscorides[2], who merely describes drugs, says that frankincense is produced in that part of Arabia called Libanotophoros, and that the best is called "Stagonias," that the Indian kind is darker, and mentions other varieties, "Syagrian," &c.

His mention of an "Indian" kind has been the source of all the controversy as to the true habitat of frankincense, for all the great Arabian writers misled themselves by this statement of Dioscorides.

The casual notices of frankincense by the Latin poets and historians are also very valuable in the present argument. It would be impossible to quote in this paper all the passages in which frankincense is named or its use implied by them; for they would simply double its bulk. The most pertinent of them are quoted by Stuckius in his "Sacrorum Sacrificiorumque Gentilium Descriptio," and in the "Hierobotanicon" of Celsius, from which two writers most of the information given in modern "Cabinets," "Treasuries," and "Cyclopædias" concerning frankincense is taken without acknowledgment, and very often without intelligence; and as Celsius makes a determined effort to exhaust all the learning on the subject, he very well proves how impossible it is to be universally learned on even so infinitesimal a subject as frankincense. In the case of some of the Latin poets (Ovid, Horace, Virgil, Martial, and Statius) the word for frankincense is always catching the eye on the look out for it. It occurs frequently in Tibullus, Claudian,[3] and Apuleius; once or twice in Plautus, Juvenal, and Lucretius; once in Persius; and nowhere, so far as I have searched, in Terence. The single quotation each from Ausonius[4] and Florus[5], which name the Lebanon as the habitat of the frankincense-tree, are quite hackneyed. It is continually occurring in Aurelius Prudentius Clemens[6], whom I have never found quoted on the subject. This mass of quotations is direct evidence of the universal use of frankincense throughout the ancient world in the worship of the gods, and of the complete communion of the nations under the Empire of Rome; such a community as steam and electricity are now once again bringing about for the whole habitable globe, and which from the irruption of the barbarians into Europe has survived alone in the spiritual domination of the Roman Catholic Church.

These passages, all taken together, prove the universal trade of the nations of antiquity for centuries in frankincense. In the way of this trade, however much its monopolists may have desired to make a mystery about it, the country from which it was procured must have become known to thousands of persons; and therefore great weight is to be given to the common consent which these passages prove that frankincense was procured from "Arabia," "the Arabians," "Sabæa," "the Sabæans," "Panchaia"[7].

As, however, in ancient times, the whole commerce of the east and west was for centuries poured into and exchanged in the coast-cities of Arabia, a commerce so rich and rare, and which so struck the imaginations of men, that some of the sublimest allusions of the Hebrew prophets are derived from it, and its fame "vibrates in the memory yet" of all the countries of the Mediterranean and Arabian seas, many products of countries further east than Arabia may have been, and indeed were, received in the west or Mediterranean countries as the products of Arabia; but, as to frankincense, it is only necessary to reply here that it is always mentioned as a foreign production in ancient Hindoo books, according to Heeren[8], and that to this day the people in the bazaars of Western India say that it comes from Arabia, and that not, I believe, because it is simply a fact of their present dealings with Arabia, but of the great myth of the ancient commerce of the Sabæans, which still lingers in the east.

The high honour in which the offering of frankincense was held, is shown by its being named as one of the three gifts of the "wise men from the east"[9], the significance of which is well illustrated by the passages in Claudian and A. P. Clemens to which I have above alluded.

In the Revelation, xviii. 11–13, we have another confirmation of the importance of the trade in it when that book was written:—"And the merchants of the earth shall weep and mourn over "her [Babylon]; for no man buyeth their merchandise any more: the merchandise of gold, and "silver, and precious stones, and of pearls, and fine linen, and purple, and silk, and scarlet, and all "thyine wood, and all manner of vessels of ivory, and all manner of vessels of most precious wood,

---

[1] Geographia Ptolemæi: Basiliæ, 1542. Orbis Antiqui tab. geograph. secundum Ptolemæum: Amstelodami, 1730. [Added in 1874. Col. Yule in a note to me of Sept. 7th writes:—"I do not suppose there is any real difference between the localities assigned "by Ptolemy and (pseudo) Arrian to the Thuriferous region, when Ptolemy's meaning, or rather his misunderstanding of his "data is understood. I have an unpublished map which Prof. Sprenger sent me last year ('*Ptolemæus Karte von Arabia Felix*') "with a letter of his, and he comes to the conclusion that Ptolemy had cut out the whole interval between Dhafar and Ras-el-"Hadd to make his coasting data fit his long sea runs. In this way Sprenger makes Syagros still Ras Fartak, but Corodanum is "both Ras-al-Hadd and Ras Mérbût, or something like that. Sprenger has taken immense pains with the geography of Arabia, "and from much recent study of Ptolemy's India I am convinced that he was quite capable of doing what is alleged. As to the "*Peratic* frankincense, surely Vincent is making difficulties. If Shehr was the Libanotophorous Region, what was more natural "than that what came from the other coast should be distinguished as 'Libanos from over the water.' If you observe the "Periphus (so far as Vincent gives it, and I have not the book itself) calls *all* the Sumali incense Peratic, none of the Arabian. "Hence it could not possibly mean 'from beyond the straits,' but may it not merely mean what is now suggested. And is it not "probable that the application of the term Shehri to the incense of *Boswellia Carterii* collected in Sumali means just that it is of "the old Shéhr quality—the real article identical with that of Shehr—the *frank* incense in fact. And this, I apprehend, is the "meaning of our own word."]

[2] Διοσκορίδης. Interprete Marcello Vergilio: Coloniæ, 1529. Et Ex Nova interpretatione, J. A. Sarraceni Lugdunæi, 1598.

[3] Miracula Christi.

[4] Monosyll. 346. 5.

[5] Lib. iii. 5. 29.

[6] Passio Eulaliæ. P. S. Vincentii Magorum munera.

[7] "I know where the Isles of Perfume are,
Many a fathom down in the sea,
To the south of sun-bright Araby!"

[8] I have it in my note book that Heeren asserts this, but have failed to verify the note for the purpose of this paper.

[9] Matt. ii. 11.

" and of brass, and iron, and marble, and cinnamon, and odours, and ointments, and *frankincense*, " and wine, and oil, and fine flour, and wheat, and beasts, and sheep, and horses, and chariots, and " slaves, and souls of men," . . . .

I believe that the Arabian writers Avicenna, Serapion, Edresi, Abulfeda, and Ibn Batuta all agree that frankincense is produced in the Hadramaut, behind Merbat and Sheba.[1] Serapion and Avicenna, misleading themselves by Dioscorides, say that it is produced in India also.[2] Abulfeda says that frankincense is found nowhere else but in Yemen. IBN BATUTA would appear to have seen the tree at Hafek or Hasik. His words are:—"Leaving Zofar (Dofar, Lee; Sephar of Bible ?— " G.B.), I proceeded to sea towards Amman, and on the second day put into the port of Hasik, " where many Arab fishermen reside. We have here the incense-tree. This tree has a thin leaf, " which, when scarified, produces a fluid like milk: this turns into gum, and is then called *loban* " or frankincense."[3]

MARCO POLO[4] writes of the city of Escier or Escher (Sheher, Saher, Sahar):—"This district " produces a large quantity of white frankincense of the best quality, which distils, drop by drop, " from a certain small tree that resembles the *fir*. The people occasionally tap the tree or pare away " the bark; and from the incision the frankincense gradually exudes, which afterwards becomes " hard. Even when an incision is not made, an exudation is perceived to take place, in conse- " quence of the excessive heat of the climate."

"The frankincense is so cheap in the country as to be purchased by the governor at the rate of ten bezants the quintal, who sells it again to the merchants for forty bezants. This is doné under the direction of the soldan of Aden, who monopolizes all that is produced in the district, and derives a large profit from the re-sale."

Of Duflar (Dofar) he writes:—"Frankincense is likewise produced here, and purchased by the " merchants."

THEVET[5] says that frankincense is produced, with myrrh, in the country about Pecher (Sheher?) and Fartack, cities of the kingdom of Aden, that there are two sorts (one, which is gathered in the dog-days, whitish, pure, clean cut, and solid, and the other, collected in the spring, which is reddish), and that the tree resembles the resiniferous *firs*. But the figure which he gives of it is of an undoubted *Boswellia*, as figures go in these old books. In the background, it is important to observe, are clumps of *firs*, not sketched on the spot, of course, like the Boswellia, but put in on the authority of Marco Polo, and the hackneyed quotations from Ausonius and Florus. And hence evidently the origin of the obstinate error that the frankincense tree is a fir.

GARCIA AB HORTO[6] says most positively that no frankincense is produced in India, and that all that is used in India, or imported into Portugal, comes from Arabia (he wonders that Dioscorides, followed by Avicenna and Serapion, should say it grows in India), that the Arabs call the darker kinds of frankincense Indian, that there are two kinds of frankincense, one that is produced on the mountains and the best, and the other, which is inferior and dark, produced in the plains, and that the tree is like the mastick,—the figure given of it in this edition being a copy of Thevet's, *without* the background of *firs*.

GERARD,[7] merely copies from Thevet.

CLUSIUS[8] merely notes on Garcia ab Horto's account of frankincense the invention of the synonym olibanum for it by Gerard in his translation of Avicenna, published at Venice A.D. 1490:— " The Arabs call it *Loban, unde Gerardus Carmonensis fecit suum Olibanum* (cap. 525, lib. ii. " Avisenæ), which is wonderful, when that chapter in the Arabic copy has not *Loban* but *Conder*." But this is not the first use of the synonym olibanum for frankincense. It is used by Pope Benedict IX. in 1033.[9]

STUCKIUS[10] says that frankincense comes from Arabia.

ANTOINE COLIN[11] merely translates the Latin epitomes of Garcia, adding an annotation which is an epitome of the statements of Thevetius, and illustrated by a copy of his figure of the frankincense-tree.

In "Purchas, His Pilgrimes" (London, 1627), NICHOLAS DOVENTON, Captain of the "Peppercorne," writes, March 1612:—"The eight and twentieth . . wee anchored in rough ground in seven " fathome, a mile and a half westward from Mount Felix" (Mount Elephant of the ancients, Ras Fieluk of the Arabs). "The nine and twentieth in the morning . . . the countrie people brought " downe to our men to sell, some store of sheep, small goats, with some small frailes of olibanum " and gumme arabick, all which they had at reasonable rates." This is the first evidence that I have discovered amongst the writers of our day of frankincense being obtained in Africa.

And Captain JOHN SARIS, in his Voyages, under March 1, 1611, writes:—"In the evening we " had ground, standing in along the land to find the bay of Feluke. . . . Here are gummes of " several sorts, very sweet in burning, also fine mats, well requested at Aden, Mocha, and the

1 Bocharti Geo. Sac. Traj. ad Rhenum, 1674; Ludg. Bat. 1692, lib. ii. c. 18.
2 Bochart, *loc. cit.*; Garcia ab Horto, Aromat. et Simpl. Hist. lib. i. c. 6, Ant. 1579.
3 Travels, translated by Lee.
4 Travels, bk. iii. ch. 40, "On the Province of Aden," Bohn's series. Also Bergeron's Collection.
5 La Cosmographie Universelle: Paris, 1575.
6 Aromatum et Simplicium Hist., Ant. 1579, *loc. supra cit.*
7 The Herbal: London, 1597.
8 Exoticarum Lib. Dec.: 1605.
9 Italia Sacra auct. Ferd. Ughello: Ven. 1717. Glossarium Manuale ad Scrip. Mediæ et Infimæ Latinitatis: Halle, 1778.
10 Sacrorum Sacrificiorumque Gentilium Descriptio: Tiguri, 1598.
11 Histoire des Drogues, Epiceries et Simples: Lyon, 1619.

" Indies; for ordinarily the Indian ships touch here both inward and outward, to buy thereof, and " of the gummes."[1]

HERBERT[2] writes of Arabia:—"But in the old time these were distinguished in to the Sabæi, " whence Guilandinus says was the queen that came to hear Solomon's wisdom, and the three Magi " who had the honour of presenting their offerings to Christ. And it is not without reason that this " part of Arabia abutting upon the Persian Gulf from against the Island of Bahrein, mentioned by " Eratosthenes, where the city of Calach was (now called Obollach), as far as Muscat, was the " Sabæan land, which, from the abundance of gold there found, was reputed Ophyr, though, indeed, " both Sabæa and Ophyr are near Ganges. From the plenty of frankincense and myrrh, it was " called *Thurifera Regio*, most abounding near the hilly country of Merbat and Segar (Sheher), " neighbouring the land of Hadramat or Atramit, as Pliny and Pomponius Mela (lib. iii. c. 18), " thus:—'*Sabœi Arabiæ felicis tenent partem ostio maris Persici proximi Carmaniæ, ubi montes* " '*Asabi sunt:*' albeit many contrarily suppose that Saba or Sheba (which Strabo, lib. xvi., calls " Metroba) was in the western part of Arabia, near the Red Sea. I take leave to digress a little " further. After the confusion of tongues, which was 120 years after the flood," &c.

Of Benjamin, he writes:—"Arabia has good, but Pegu and Siam better."

SALMASIUS[3] says that the word olibanum is from the Greek ὀλίβανος, and sums up the statements of Theophrastus, Pliny, and the Arabs in favour of Arabia being the land of frankincense.

BOCHART[4] says that frankincense is not obtained from India, or Africa, or Panchaia, or Mount Lebanon, but solely from Saba. Both sacred and profane writers testify to this. The thuriferous region lies between the Sachalitic Gulf and Syagrian promontory; and Cana (Makalla) is the emporium of the trade.

OVINGTON[5] writes:—"Arabia Felix . . . abounds with several kinds of drugs, with balsam and myrrh, incense, cassia, manna, dates, gold, frankincense, and pearl."

"Dofar . . produces only some olibanum, coco-nuts, and butter."

"Seer (Sheher) . . . is much frequented by ships from several ports, viz., Muscatt, Gombroon, Suratt, and Gella (Zeyla), and some other places on the Abasseen shore, from whence they bring butter, myrrh, and slaves; and those from Muscatt and Suratt transport with them olibanum, and aloes, and what the port affords."

Aden: "they carry nothing from hence but coffee, aloes, olibanum, and myrrh; the last three of which are the products of the country."

Mocha exports "coffee, senna, and some aloes hepatica:" imports "from Casseen [Keshin, near " Ras Fartak?] olibanum," "from Socotra aloes," and "from Gella [Zeyla] and other parts of " Abasseen myrrh, and from Socachim [Suakin?] elephants' teeth and gold."

HAMILTON[6] writes:—"Cassin [Keshin, near Ras Fartak] that lies under the meridian of Cape " Guardafoy, and under the prodigious high mountains of Megiddo, on the coast of Arabia." "The " product of the country (besides common roots, and fruits, and animals) is myrrh and olibanum " or frankincense, which they barter for coarse calicos from India; but they have no great commerce " with strangers."

"Nor has Doffar any better trade."

He mentions Mount Felix under the name of "Baba-fileck," or "the Camel's Hump."

POMET[7] is supposed to assert that frankincense is a product of Mount Lebanon; and he indeed says this of male frankincense; but in a separate chapter on Mocha frankincense he says that it is brought to Europe by the East India Company, and so called Indian frankincense.

CHARDIN[8] says "that the frankincense-tree, which resembles a large pear-tree, grows in Persia, " particularly in Caramania, on the mountains. You find also there the terebinth-tree, the almond, " and the wild chestnut."

CELSIUS, in his famous "Hierobotanicon," comes to the conclusion that Arabia is the native country of frankincense. He quotes from Fragosus:—"It is most true that frankincense does not " grow in any part of India, since whatever frankincense is consumed in India, and whatever is " conveyed thence to Portugal, is brought out of Arabia, where alone it grows, and is called Lovan. " On which account, I believe that writers are mistaken, who consider that it is called Libanum in " Greek, and olibanum in commerce, from the fact of its growing in Mount Lebanon."[9]

LOBAT[10] says that olibanum is not only a production of Arabia, India, and of Mount Lebanon, but of Western Africa.

LINNÆUS referred (I take this statement at second hand) frankincense to an unascertained juniper; and scientific botanists after him boldly specified *Juniperus Lycia* as the tree; and, in fact, until a very few years ago (1832) *Juniperus Lycia* was generally held to be the frankincense-tree.

---

1 This passage is referred to by Vincent, "Commerce and Navigation of the Ancients," Lond. 1807, vol. i. p. 90.
2 Travels: London, 1677.
3 Plinianæ exercitationes et de Homonymis Hyles Iatricæ: Traj ad Rhenum, 1689.
4 Geographia sacra: Traj. ad Rhenum, 1674, and Ludg. Bat. 1692.
5 Voyage to Suratt (1689): London, 1698.
6 Account of the East Indies, 1688–1723, in Pinkerton's "Travels."
7 Histoire des Drogues: Paris, 1694; translation, London, 1737.
8 Voyage en Perse et l'Orient: Amsterdam, 1711.
9 Olavi Celsii Hierobotanicon, sive de Plantis Sacræ Scripturæ Dissertationes breves: Amstelodami, 1748, pp. 231–246.
10 Nouvelle Relation d'Afrique: Paris, 1728.

NIEBUHR is always cited as an authority in this controversy; and the statement attributed to him is:—"We could learn nothing of the tree from which incense distils, and Mr. Forskal does "not mention it." But this is not what Niebuhr says, but only Heron's (Edinburgh, 1792) very free and easy translation of what he says. But even Heron's translation continues, "I know that "it is to be found in a part of Hadramaut, where it is called oliban. But the Arabians hold their "own incense in no estimation, and make use only of that which comes from India. Probably "Arabian incense was so called amongst the ancients, because the Arabians traded in it, and "conveyed it from India to the ports of Egypt and Syria."

Now, what Niebuhr says in the "Description de l' Arabie," Copenhagen, 1773, is, in epitome, this:—Anciently Arabia was not less celebrated for its incense than for its gold; but all the incense received from Arabia was not produced there. Arrian and others have shown that much was imported from Abyssinia and India, and from beyond it. Indeed it is chiefly cultivated in Arabia, about Merbat, Schähr (Sheher), and the kind called *Liban* by the Arabs, and frankincense by the English, which is the worst of all. The Arabs get other kinds from Abyssinia, Sumatra, Siam, Java, and amongst these what they call Bacher Java, and the English benzoin, and the poorest sort of which is esteemed better than the best frankincense. One may conclude from this, that many kinds passed in ancient times under the names of Arabian incense, although from countries far beyond it. In this way Arabian coffee is called, in Europe, Levant coffee. The Arabs do not set much on their olibanum; for they import the incense of India ["l'encens des Indes" (benzoin ?)] and the mastic of Scio (p. 126).

Moses, he testifies, in Genesis, names the cities of Yemen and Hadramaut, as if he had personally known them. Hadramaut is the Hatzarmaveth of Moses, which signifies, in Hebrew, the Court of Death, and, in Arabic, the Region of Death. In the Hadramaut, Dafar, Merbat, and Hafeck export the best olibanum (pp. 245–255).

BRUCE, in the Atlas to his Travels (Edinburgh, 1805), figures a plant which he says is called *Angouah* by the people about the Tacazze, and believed by the Abyssinians to yield true frankincense; and he adds, "in reality it produces a gum much resembling it." It is undoubtedly the *Boswellia papyrifera* of Richard, and the first accurate figure ever drawn of a frankincense tree. But botanists seem never to have given Bruce the credit of the discovery of this plant, and possibly because he has figured as its fruit the capsule of an undoubted *Bombax*. Stackhouse[1] lays no stress on the plant, merely observing, "Libanus vel Libanotos Theophrasti. This, according to "Sprengel, is the *Amyris Kataf* of Linnæus [Forskål ?]. It is to be regretted that the description "of Theophrastus is so incorrect as not to ascertain the point." Royle[2] merely remarks that the *B. serrata* of Roxburgh is not Bruce's plant. Bruce observes that the frankincense land stretches from Abyssinia away to Cape Gardafui[3].

COLEBROOKE, in 1807[4], proved that an exudation allied to frankincense was yielded by the *Boswellia serrata* of Roxburgh, now known under the name of *B. thurifera* Colebrooke. As soon as Colebrooke's discoveries became known, it was concluded that this was the olibanum of commerce, and it was even denied that Arabia yielded any olibanum. But I will not anticipate.

DELILE[5], in 1820, described a plant called *Kafal* by the Arabs of Fazogl, *Galgalaan* by the negroes of the country, and *Loban-adan-culân* in Sennar, and named it *Amyris papyrifera*.

In MILBURN's admirable work on *Oriental Commerce*, published in 1825, it is stated that *Jedda*, *Aden*, and *Massowah* import benjamin, and *Zeyla* and *Berbera* export olibanum, and that olibanum is procurable in the Persian Gulf.

Of Berbera he writes that "a fair is held here from October until April, and the caravans arriving "*from the interior* during that period bring large quantities of gum-arabic and myrrh. Olibanum "is chiefly produced on the coast, between 'Barbora,' or 'Burbureea,' and Cape Gardafui, and "exported in Arab vessels from a small port near Cape Felix [Mount Elephant of the ancients; "Ras Fieluk; Cape Felix, or Felles, of the Portuguese; Jibbel Feel, Capt. Saris; Baba-fileck, "Hamilton, *suprà*]. A small proportion of these articles reaches Bombay and Europe; the largest "part goes up the Red Sea to Egypt."

Amongst the articles to be obtained at Calcutta and Bombay olibanum is not mentioned by him.

ENDLICHER, in 1838, described and figured his *Plösslea floribunda*[6], obtained from Fazogl, and which Royle was one of the first to recognize as a *Boswellia*, and which is the same plant as Bruce's "*Angouah*" and Delile's *Amyris papyrifera*.

WELLSTEAD[7], in 1838, and Johnston[8], in 1842–43, both stated that large quantities of olibanum are exported from the Soumali coast. Wellstead says, "two kinds of frankincense are brought here "for exportation to Hindostan, one called *Luban*, from Hadramaut, which is a powerful aromatic, "used in the temples and houses for fumigation; the other *Lubau mati*, less fragrant, but preferred "for chewing." Again, "In the mountains behind Saher and Makallah are found dragon's blood [Bdellium ?] and frankincense trees. I have often seen the former, but never the latter."

HAINES[9] also did not find the frankincense-tree in his examination of the Arabian coast "commencing from the entrance of the Red Sea, and continuing so far as Messenaad, in latitude

---

1 Extracts from Bruce's Travels: Bath, 1815.

2 Pharm. Journal, v. 541: 1845.

3 Travels, vol. i. p. 356.

4 Asiatic Researches, vol. ix. Calcutta, 1807; Linn. Trans. vol. xv.

5 F. Caillaud, Voyage à Méroë: Paris, 1820, vol. ii.

6 Iconographia, 119–120; Nov. Stirp. Dec. n. 47.

7 Travels in Arabia: Lond. 1838.

8 Travels in Southern Abyssinia: Lond. 1842–43.

9 Transactions of the Bombay Geo. Soc. xi. [Added 1874. Col. Yule in a note to me of Sept. 7th writes:—"I find that "poor old Haines *did* see, with his own eyes, and speak of olibanum trees. See Journal Royal Geo. Soc. for 1845, vol. xv. "pp. 118, 119."]

"15° 13′ N. and longitude 53° 43′ 25″ E.," and he says that the frankincense exported from Makalla is imported from Berbera.

CRUTTENDEN[1], in 1837, saw the frankincense tree of Arabia, on his journey from Merbat to Dyreez, the principal town of Dejar, and in 1843 the frankincense tree of the Soumali country.[2] He is reporting on the Mijjertheyn tribe of Soumalis, and says:—"During the hot season, the men "and boys are daily employed in collecting gums, which process is carried on as follows:—About "the end of February, or beginning of March, the Bedouins visit all the trees in succession, and "make a deep incision in each, peeling off a narrow strip of bark for about five inches below the "wound. This is left for a month, when a fresh incision is made in the same place, but deeper. "A third month elapses, and the operation is again repeated, after which the gum is supposed to "have attained a proper degree of consistency. The mountain sides are immediately covered with "parties of men and boys, who scrape off the large clear globules into a basket, whilst the inferior "quantity that has run down the tree is packed separately. The gum when first taken from the "tree is very soft, but hardens quickly. The flame is clear and brilliant; and the traveller is fre-"quently amused by seeing a miserable Bedowi family, cowering under a wretched hovel, eating "their scanty meal by the light of half-a-dozen frankincense-torches[3]. Every fortnight the mountains "are visited in this manner, the trees producing larger quantities as the season advances, until "the middle of September, when the first shower of rain puts a close to the gathering of that "year."

The export, he goes on to say, is in the hands of "those never-failing speculators, the Banians of "Porebunder (Kattiawar) and Bombay." At the close of the north-east monsoon they arrive at the coast, and settle at Feluk (Mount Elephant of the ancients, Cape Felix of the Portuguese, and Jibbel Feel of Capt. Saris), Bunder Marayah, and Bunder Khor, and other Bunders. "The Bedowins "from the interior immediately visit them; and as there is no one to compete with them, they "manage to engross the greater part of the trade." He estimates that the quantity annually exported is:—

	Tons.
To Bombay - - - - - -	377
To Red Sea - - - - - -	235
To Arabia - - - - - -	120
Total - - -	732

"The trees that produce *Laban* or frankincense are of two kinds, *Luban Meyti* and *Luban Bedowi*. Of these the *Meyti*, which grows out of the naked rock, is the most valuable, and, when clean packed and of good quality, it is sold by the merchants on the coast for 1¼ dollar per frasila[4] of twenty pounds. The *Luban Bedowi* of the best quality is sold for 1 dollar per frasila; of both kinds the palest colour is preferred. The trees vary greatly in height; but I never saw one above twenty feet, with a stem of nine inches in diameter. Their form is very graceful, and when springing from a mass of marble on the brink of a precipice, their appearance is especially picturesque."

He mentions also that myrrh is chiefly brought from Murrayhan in the interior, and sold at Bunder Murraayha, near where he also saw a few myrrh-trees.

In 1843, also, KEMPTHORNE and VAUGHAN made a short journey into the Soumali country, "from Bunder Marayah, the chief town of the Mijjertheyn Soumalis, near Ras-Feeluk" (Mount Elephant, ancients; Cape Felix, Portugese; and Jibbel Feel, Saris). Kempthorne's description of the frankincense-tree, which he saw here on the limestone mountains, three miles inland from Bunder Marayah, I have copied after my botanical description of it below[5]. Royle says[6] that, "Bennet identified this plant (which is Cruttenden's *Luban Meyti* also) with Endlicher's *Plösslea* "*floribunda*, but on an examination of the bark only." Kempthorne says, "During the south-west "monsoon, the pastoral tribes in the neighbourhood of Ras Feluk collect large quantities of "frankincense, which they barter to the Banians, of whom a few reside in the villages along the "Abyssinian coast. Boats from Maculla come and fetch the gums, which are exchanged for a "coarse kind of cotton cloth worn by the Soumalis." And again, "I think that the localities from "which the ancients obtained their principal supply of frankincense were situated on the almost "inaccessible mountains of Adel, in Africa, facing the southern coast of Arabia; for the trees as we "saw them, extended over such a vast tract of land that large quantities of the gum even now "must be collected annually, as in the days of the Greeks and Romans. The plant grows at the "height of between 2,000 and 3,000 feet above the sea." He mentions that Cruttenden had before visited the country, but that Vaughan and he were the first Europeans who had ever visited the frankincense forests of Adel. Vaughan states[7], "The Luban-tree is a native of the eastern coast of "Africa, and flourishes on the highlands which intersect the whole of the Soumali country, where "I had an opportunity of seeing it in 1843, not far from Cape Gardafui." And he gives the following as the different kinds of Luban found by him in the bazaar of Aden.

Luban maitee, from Bunder Mait, collected chiefly by the Abardagahala Soumalis. *Luban nankur* or *aungure*, from Bunder Aungure and the country of Door Mahomed and the Abardagahala

---

[1] Trans. Bombay Geo. Soc. i.

[2] Trans. Bombay Geo. Soc. vii. 1846.

[3] Compare Pliny, *suprà*.

[4] Compare Purchas, *loc. suprà cit.*

[5] See Trans. Bombay Geo. Soc. vi. and xiii.; Harris, Highlands of Æthiopia, vol. i. 417, 1844; Pharm. Journal, iv. 1844. [Added 1874. Col. Yule in his note before quoted further observes:—"Also you put Kempthorne seeing them (the frankincense "trees) in Soumali land in 1843. But from Ritter xii. 360, who quotes proceedings Bombay Geo. Soc. May 1838, p. 55, it would "seem that Kempthorne must have reported on them in 1837 at latest."]

[6] Pharm. Journal, v. 1845.

[7] Pharm. Journ. vol. xii. 1853.

Soumalis. *Luban makur*, from the ports of Ras Rurree, Khor Bunder, Alholu, and Bunder Maryah, and Bunder Khassoom, in the country of the Wursangali and Mijjerthên Soumalis, about Cape Gardafui. This drug is collected in March, April, and May, and chiefly finds its way to Bombay through the *entrepôts* of Maculla and Shehr. *Luban berbera* or *muslika*, collected by the Aial Yunus and Aial Hamed Soumalis, and which is in agglomerated *dark, vitreous tears*, and largely used by the Arabs themselves; and *Luban Marbat*, or *Saharee Luban*, the frankincense of the libanotophorous region of Arabia, the highest-priced of all, and exported in enormous quantities to Bombay.

RICHARD[1], in 1845, described and figured his *Boswellia papyrifera*, identifying it with Delile's and Endlicher's plants, which are identical with Bruce's *Angouah*. The Abyssinian names he states, on the authority of Dillon and Schimper, to be *Makker*. "It grows on the mountains which encase the Tacazze."

In 1847, CARTER published[2] a figure and description of the frankincense-tree of Arabia, with remarks on the misplacement of the libanotophorous region in Ptolemy's geography, a paper which of course decided the vexed question whether Arabia produced frankincense or not, and for the first time accurately fixed the position of the Arabian thuriferous region, but which actually remained unnoticed, until I drew attention to it a few years ago, as will be presently told. He placed it under Colebrooke's species; and hence probably, in part, the reason why it was overlooked; but it was a distinct technical species. It was desirable to find out where his original specimens were; and on my inquiry he wrote to me, on the 1st of November last, that—

"The original drawings of the Arabian frankincense-tree I sent in to the Government of Bombay, for transmission to the Court of Directors, among seventeen original sketches, and eleven specimens of gums, my gatherings in the south-east coast of Arabia, with short descriptions of each, in April 1848; and, on the 28th of February 1851, I got the following reply through the Bombay Government, extracted from a dispatch of the Court of Directors:—'Great credit is due to Assistant-surgeon Carter,' &c.

"It was from the original sketch that I sent home to the Court that I *traced* the lithograph of the same, which appears with my little memoir on it. But I sent home delicately coloured drawings also."

I applied at the India House, to be shown the specimens and drawings, but was informed that they did not appear to be any longer there.

My next question of Carter was, if he had himself seen the Arabian frankincense-tree *in situ*, or if merely specimens had been collected for him by the Arabs.

His reply, January 28th last [1869], was: "On the 25th December 1844, I made my first " excursion in the south-east coast of Arabia. It was to see the frankincense-trees growing at the " foot of the great limestone cliffs, near the village of Merbat. They were then leafless; but the " odour of the gum-resin was convincing, even before they were touched. The same was the case " with *Boswellia serrata*, which covers the sides of the Ajunta range, which I traversed with Bháu " Daji. The whole air was redolent with their fragrance. On the 21st of April 1846, I found " the frankincense-tree growing on the *detritus* of the great white limestone cliffs of Ras Fartak, " within a few feet of the sea. It was then in full foliage but without flowers. On the 22nd " of May following, a little further to the north-east, I got a Bedoo to go up into the limestone " mountains, close to the shore, and bring me some branches in flower. He brought them, with " some rooted stems also, which I brought to Bombay, and from which the plant in the Old Gardens " at Bombay was obtained. Having specimens of the branches and leaves, which I had myself " gathered just a month before, from the trees under Ras Fartak, I compared all, and found all " to belong to the same species, and forthwith made the drawings on board, of which you have the " lithograph."

Carter, after identifying his plants with Colebrooke's—with a query, however, and a doubt still further expressed in a foot-note—observes, "In addition to India and that part of Arabia which I " shall presently point out, the frankincense-tree is found in great abundance in Eastern Africa, " on the limestone mountains which extend westward from Cape Gardafui through the country of " the Soumalis. I have seen a living specimen in foliage brought from thence, and large quantities " of the gum, which is imported at Makalla for re-exportation to India. Both the produce of the " tree of Africa and Arabia appear to be the same."

"Ibn Batuta calls the tree *al Kundooroo*. The gum is called by the Arabs *Laban*. The Maharas call the tree *Maghrayt d'sheehaz*, the gum *sheehaz*.

He then epitomizes the account given by Theophrastus of the frankincense hills and the tree and the manner of collecting its gum-resin, adding, "but for the disappearance of the Sabæans and " their towns, I could not offer a more accurate description."

"The gum is procured by making longitudinal incisions through the bark, in the months of May and December . . . On its first appearance, the gum comes forth white as milk, and, according to its degree of fluidity, finds its way to the ground, or concretes on the branch, ... from whence it is collected by men and boys employed to look after the trees by the different families who possess the land in which they grow."

He found the *Moql* or bdellium-tree growing with the frankincense, but not the myrrh-tree, as Theophrastus relates.

---

[1] Voyage en Abyssinie, pendant 1839–1843, par M. Théophile Lefebre, &c., vols. 4 & 5; Tentamen Floræ Abyssinicæ, auctore Achille Richard, Atlas, tab. 38.

[2] Journal of the Bombay Branch Royal Asiatic Society, vol. ii.

"The insalubrity of that part of Arabia . . . is fabulous . . . It is most invigorating and healthy, the abode of the Bedouins, and the resort of the lowland people of the coast during the hot season."

Of the thuriferous district he writes:—"Coming from the north-east, we first meet with the " frankincense-tree on the Sabhan mountains, in latitude 17° 30′ north, and longitude 55° 23′ east " [near Merbat ?], where the desert ends, and the wooded mountainous region commences; and fol- " lowing the coast, which runs south-west, we find the frankincense exported from the different " towns gradually diminishes, after the bay of Al Kammar, until we arrive at Makalla, from whence " none is exported from the interior of Arabia, and but little used, except what is brought from the " African coast, opposite that town. By the same inquiry we learn that the produce of the Arabian " trees is exported in largest quantities from places on that part of the coast which intervene the " longitude and latitude mentioned (viz. 17° 30′ north, and 55° 23′ east) and the town of Damkote, " in the bay of Al Kammar, 52° 47′ east longitude [Ras Fartak ?]. Between these two points the " trees are congregated in two distinct localities—on the summits and sides of the highest range of " mountains near the coast, and on the plain between them and the sea; the former is called *Nedjee*, " or high land, and the latter *Sahil*, or plains on the coast. The *Nedjee* is about two days' " journey from the shore; it is the most elevated portion of the great limestone formation of this " coast, which, from a height of 5,000 feet, here descends in sudden and lofty steps upon the Arabian " Sea. To get to it you first cross the *Sahil*, already mentioned, and then ascend a minor range, " which is covered with long grass and trees, and, after passing a less fertile region called the *Găthăn*, " at last arrive at the *Nedjee*, where there is no grass, and but few trees besides those which produce " the frankincense. The soil is red and sub-argillaceous; and in consequence of its scarcity, the " trees are generally found growing out of the crevices of the limestone-rock. It is from this part " that the frankincense is chiefly brought, and, as I have before said, that the largest quantities of " it are exported from the different towns on the coast, between the longitudes of 52° 47′ and " 55° 23′ east." "The libanophorous region therefore lies behind the towns on this part of the " coast, where Theophrastus places it, and not as Ptolemy places it in Oman." I think this is hypercritical on Ptolemy; but it is invaluable as accurately limiting the frankincense-region of Arabia and demonstrating the extreme accuracy of Theophrastus.

That Haines did not find the frankincense-tree on the Arabian coast, between Aden and Makalla, confirms the accuracy of both Theophrastus and Carter.

When I was placed in charge of the Agri-Horticultural Society's "Old Gardens" in Bombay, in 1859, I found the tree which Carter had brought from Arabia growing there, and labelled, by Stocks, *Boswellia papyrifera;* and in my catalogue of the Government Central Museum, published in 1862, I adopted *Boswellia papyrifera*, Richard, which Royle had called *Boswellia floribunda*, as the source of olibanum; for I had ascertained that *Boswellia thurifera*, Colebrooke, produced none of the olibanum of commerce, but only an oleo-resinous exudation. But I had not seen Endlicher's and Richard's descriptions and figures, nor probably had Stocks: and I therefore asked Colonel (then Captain) Playfair, at Aden, whether he could procure me any cuttings of the African frankincense-tree. Colonel Playfair sent me a large collection of several varieties of dried leaves and of cuttings, accurately labelled with their native names, with samples of the kinds of frankincense which they respectively yielded. At my request, he subsequently sent duplicates of his dried specimens to Kew. On the cuttings which were sent to me striking and leafing, I found that I had three kinds:—*Yegaar*, yielding *Luban maitee*, an undoubtedly new species, and *Mohr Add* and *Mohr Madow*, yielding *Luban sheheri*, the bulk of the olibanum exported from the Soumali country. Judging by the young leaves solely, I should have been inclined to consider Colebrooke's, Endlicher's, Richard's, and Carter's plants, and Roxburgh's *Boswellia glabra* all only more or less variations of one species, and *Yegaar* the only second species. Indeed, if I followed my own erring inclination, I should hold so now.

Amongst Playfair's dried specimens of *Mohr Add* and *Mohr Madow*, were the leaves of a variety of each, which he said yielded an inferior olibanum, called *Luban bedowi*.

Carter's plant never flowered with me, nor until last year would Playfair's. But he had planted duplicates of the cuttings he sent to me, in the romantic little gardens laid out by him near the celebrated tanks of Aden. They had not flowered before he left Aden for Zanzibar; but when passing Aden, in June 1867, he found that one of the plants had flowered, and he sent its flowers to Kew. When I saw these in July 1867, it was not known to which kind they belonged, as Playfair had not picked any leaves with them. I therefore, on my way back to Bombay, in November 1867, visited the gardens, and found from the gardener that it was *Mohr Madow* that had flowered the previous June. Seeing the plants were leafing too much, I left directions to water them less abundantly than they had been, and to send me any flowers that they produced. On reaching Bombay, I also stopped watering the plants in Victoria Gardens, the Agri-Horticultural Society's new Gardens in Bombay; and early in the monsoon of last year *Mohr Add* flowered. In September last, I was again at Aden, and then found *Yegaar* in full bloom. It is from these materials that I have, with Dr. Thomson and Professor Oliver's kind assistance, prepared the descriptions, and Mr. Fitch has drawn the figures, of the three species which I have named:—

Boswellia Carterii (Mohr Madow).
Boswellia Bhau-Dajiana (Mohr Add).
Boswellia Frereana (Yegaar).

In rearranging the genus I have also included Roxburgh's *B. glabra* under Colebrooke's *B. thurifera*.

Mohr Madow I have named as if Carter's plant; that is, as the specimens of his plant have disappeared from the India House, I have described it from my specimens of *Mohr Madow*. If I did not name *Mohr Madow* as Carter's plant, certain as it is that it is the same species as his Arabian plant, on this being indisputably determined by the receipt of new specimens, it would become absorbed in *Mohr Madow* under its systematic name, and Carter's connexion with his own discovery would be lost sight of[1].

Mohr Add is named after Dr. Bhau Daji, the distinguished Hindu physician of Bombay, well known in Europe also as an oriental scholar, and whose eminent services to his countrymen and city Sir Seymour Fitzgerald has most honourably recognised in making him sheriff of Bombay for the present year.

Yegaar I have named after Mr. W. E. Frere.

In 1867, Mr. Hanbury showed me a most interesting article on frankincense, in the "Schweizerische Wochenschrift für Pharmacie" of 13th of May 1864, by Dr. Flückiger of Bern, who quotes a large number of authorities, especially German, on this controversy. He had had Carter's plant brought to his notice by Hanbury; and he concludes that it is the same plant as *B. papyrifera* of Abyssinia, and proposes in consequence to include both under the name of *B. sacra*. But Carter's plant is distinct from *B. papyrifera;* and moreover it is now plain that more than one species of African olibanum yields the olibanum of commerce, whereas Dr. Flückiger prematurely includes them all under *B. papyrifera*, a species not known to supply any of the olibanum of commerce. He had heard of the facts ascertained by Colonel Playfair and myself regarding the thuriferous Boswellias of the Soumali country, but only orally, through a mutual friend of his and Carter's, and hence, I believe, the error into which he is led.

Conclusions.

It will, I believe, be now agreed that Arabia produces frankincense, and in the very region generally indicated by the common opinion of the ancient world and so accurately limited by Theophrastus. Now that it has been demonstrated by Carter, that the libanotophorous region lies along the coast of Hadramaut, the agreement between the statements of the Bible, Herodotus, Theophrastus, Diodorus Siculus, Strabo, Pliny, Arrian, Ptolemy, the Arabs, Marco Polo, Bochart, and Celsius is very striking.

It will, I believe, be agreed also, as implied in the more ancient references to Arabia, and expressly asserted by Diodorus, Strabo, Arrian, and others of the ancients, and demonstrated with such fulness by Cruttenden, Kempthorne, and Vaughan, and the overwhelming evidence of modern travellers (beginning with Bruce) generally, that the Soumali country also produces frankincense, and probably the bulk of the olibanum of commerce. The agreement as to the region about Mount Elephant (Ras Fieluk, Cape Felix, Jibbel Feel), between Arrian, Capt. Doventon, and Capt. Saris in Purchas, Milburn, Cruttenden, Kempthorne, and Vaughan, is indeed quite startling.

It will be admitted also, I believe, that *Boswellia Carterii* (*Mohr Madow*), which with *B. Bhau-Dajiana* (*Mohr Add*), the source of the bulk of the olibanum of commerce exported from the Soumali country, is the same species as the plant (*Maghrayt d'sheehaz*) which Carter found produced the frankincense of the Hadramaut—and also that *Boswellia thurifera* (including *B. glabra*) of India, and *B. papyrifera* of Abyssinia, whilst thuriferous species, are not known to yield any of the olibanum of commerce.

It is very surprising that so great weight of evidence in favour of frankincense being produced in Arabia Felix and the Soumali country should ever have been set aside for the idle fancy that India was the source of the olibanum of commerce. Twenty years ago it was clearly demonstrated that Arabia produced it, and one at least of the plants which produced it was fully described and figured. The error has proved as inveterate as the fiction that frankincense was yielded by *Juniperus Lycia*. Colebrooke's discovery was in 1807. In volume i. of Woodville and Hooker's "Medical Botany," London, 1832, olibanum is still said to be derived from *Juniperus Lycia;* and it is added, "it is brought from Turkey and the East Indies, but that which comes from India is "less esteemed." But in vol. ii. *B. serrata*, Rox. (*B. thurifera*, Colebrooke), is adopted as its source, and it is added, in reference to the new view, that olibanum is obtained from India, "but "it is also brought from the Levant in casks and chests."

Fleming[2], in 1810, wrote, "That naturalists should have remained in ignorance and in errour, "until almost the present day, respecting the tree which yields a substance so long known, and so "universally used, must appear not a little surprising. Such, however, is the fact, and the merit "of having discovered the true origin of this celebrated incense, is due to Mr. Colebrooke, who "has ascertained and proved most satisfactorily, that the olibanum, or frankincense of the ancients, "is not the gum-resin of the *Juniperus Lycia*, as was generally supposed, but the produce of our "*Boswellia thurifera*."

In Waterson's "Cyclopædia of Commerce," in 1847, it is stated, "Olibanum comes from India. "An Arabian kind formerly imported from the Levant is now seldom met with, and its origin "is a matter of doubt."

Pereira, in the editions of his "Materia Medica" previous to 1857, says that "olibanum is "imported from India, in chests, and that African or Arabian olibanum is rarely met with in this "country." But in the edition of 1857, by Taylor and Rees, it is said, "Strictly speaking, there "does not seem to be any Indian variety [*i.e.* commercial?] of this gum, but, like acacia and

---

[1] Carter's (the Arabian) plant has flowered since this paper was read; and a figure of it, from the flowers kindly sent to me by Mr. Wellington Gray, B.M., the present Director of the Victoria Gardens, Bombay, is now added, after that of the Soumali *B. Carterii* or *Mohr Madow*.

[2] Asiatic Researches, vol. xi.

" myrrh, it reaches Bombay from the Persian Gulf." Royle, in all the editions of his "Manual of Materia Medica," even in the posthumous edition of 1868, edited by Headland, says that there are two kinds of olibanum:—"Indian, which is the most esteemed," and "is imported in chests " chiefly from Bombay; and African, which is no doubt produced on the hills of the Soumali coast " westward, from Cape Guardafui, and carried to the Arabian coast, chiefly by native boats from " Moculla." In Maunder's "Treasury of Botany," London, 1866, edited by Lindley and Moore, it is written, "*B. thurifera* furnishes the gum-resin known as olibanum, which is supposed to have " been the frankincense of the ancients. African olibanum, a drug rarely met with in this country, " has been conjectured with much probability to be the product of *B. papyrifera*." But the most surprising surrender to this false summons is that of the profession of interpreters of the Bible. Tristram, in his "Natural History of the Bible," London, 1867, writes, "Frankincense, the fragrant " gum of an Indian tree, procured through Arabia. . . . though brought from Sheba, in Arabia, " at a very early date. . . . yet we have no reason to believe the frankincense-tree grew in " Arabia; at least it cannot be traced there now: and, as in the case of cassia and other spices, " the Arabian merchants were not likely to be communicative in the secrets of their monopoly [!]. " To the present day the best frankincense used in Turkey is imported through Arabia, from the " Eastern Archipelago. It is also largely imported into Europe from Bombay, for use in the " Romish and Greek ceremonials, but of inferior quality."

It is, I think, probable that other odoriferous gum-resins, oleo-resins, and balsams may have been confused with frankincense by ancient writers, as by modern ones. Niebuhr expressly confuses benjamin with frankincense, as does also Herbert before him; and this confusion, I believe, underlies the whole error of India being the source of the olibanum of commerce. Milburn and others confuse it with the mastic, which is irregularly imported in very large quantities into Bombay. It has been popularly confused with balm:—

"Let *India* boast her plants, nor envy we
The weeping amber, or *the balmy tree*."[1]

[(*Added* Dec. 3.) So Mrs. Manning, in her "Ancient India," vol. ii. p. 348, says "balm and myrrh " may or may not have come from India."] And the predisposing cause of all this confusion has been the unfounded belief that Dioscorides says that frankincense comes from India. It is remarkable that the benzoin of Java is said not to be mentioned by any ancient writer; and yet it is inconceivable that such a fragrant balsam should not have been known to them, and I cannot help suspecting that benzoin at least was largely used as frankincense in ancient times. It is used throughout Japan, China, Burmah, Siam, India, and Arabia, at this day, with frankincense. Books say that frankincense is still the sweet perfume "of our catholic churches;" but Stackhouse says that it is not now used, benjamin being substituted for it, an assertion which every inquiry of my own contradicts. In the Malayan language the same word is used for frankincense and benzoin.[2] Dioscorides, also, as has been shown, distinguishes an Indian frankincense; and so, when Colebrooke's discovery came, it was accepted as settling the controversy. But I question whether some variety of the African frankincense was not the Indian frankincense of Dioscorides.

I have given a few extracts above, showing the change in the trade-routes between Europe and the East; and these changes also help to explain how it has come to be believed that India produces the olibanum of commerce. Errors from such changes are very common in connexion with commercial articles; calumba, East-Indian senna, and balsam of Peru are illustrative instances. The similarity between the Hebrew, Arabic, and Greek names for frankincense and Mount Lebanon probably led to the error of Ausonius, Florus, Pomet, and others, that Mount Lebanon produced frankincense. Celcius quotes an Orphic verse and passages from Euripides and Athenæus which also assert or imply that frankincense was a product of Syria. And the error seems to survive in the nations of the Mediterranean shores; for at the Universal Exhibition of Paris, in 1867, I saw several bottles of frankincense labelled "olibanum, so called because it comes from Mount Lebanon." I also saw there aconite, labelled "Aconitum Nepaullus, so named because its root is the famous " poison of Nepaul."

The "Oxford Encyclopædia" says that olibanum is *quasi* Oleum Libani, being distilled from the bark of a tree growing on Mount Lebanon!

It would be presumption in me to enter into the etymology of the synonyms of frankincense I have been unable even to discover when the English word frankincense came first to be used.[3] But I cannot help asking if *thus* and *tus* are really the same word, why the Italian for frankincense should be *olibano*, as if formed from the Greek or Arabic, if there be any connexion, and what its significance may be, between the Arabic words "Cunder," "Al-Kundooroo," and Sanscrit "Cunduru," for frankincense, and the Greek Χόνδρος, and why the Bay of Al Kammar is so named.

---

[1] " . . . . lacrymatas cortice myrrhas."—*Ovid*. "The Bee and the Orange-Tree," Fairy Tales, D'Aulnoy, trans. Planché: Lond. 1855.

[2] Marsden's Dict. of the Malayan Language; Crawford's Grammar and Dict. of the Malay Language.

[3] Minsheu, London, 1599, has "frankincense—encensio." Cotgrave, London, 1632, has "frankincense—encens. F. in drops—*oliban*." Steph. Skinner, M.D., Etymol. Ling. Angl., London, 1671, has "frankincense, thus, *q. d.* Incensum, *i. e.* Thus libere seu " liberaliter, ut in sacris officiis par est, adolendum." Rob. Brady, M.D., "An Introduction to the Old English History and " Glossary." London, 1684, has no mention of frankincense. Gower, in his story of Leucothea, from Ovid's "Methamor," turns her "into a flour was named golde" (Confessio Amantis, Pauli, ed. Lond. 1857, lib. v.). [Added 1874. "Col. Yule, in a note to " me of Sept. 7th, writes:—Franc in one of its definitions by Littré = genuine. '*Drogue franche*, qui a les qualités requises sans " 'mélange.' And I find also in Raynouard's Romance Dictionary, 'Franc = vrai, veritable." So frank incense = genuine, true, " incense, probably, and not that it was freely, lavishly, offered to the gods, as I have somewhere read, and ever since taken for " granted."]

Burton[1] mentions that "*Lubban*" is the Soumali for milk, and a term of reproach. It will have been observed that the names of the different varieties of Soumali frankincense are identical with the names of the Bunders from which they are exported. Have the Bunders given their names to the varieties of frankincense, or these to the Bunders? Cruttenden says that the myrrh of the Soumalis comes from a place in the interior of their country called Murrayhan, and is exported from Bunder Murrayha—with which name no variety of frankincense corresponds. Does this indicate any etymological connexion with myrrh, the links of which the Soumalis themselves have lost? Also, what is the derivation of the name of the Soumali port of Berbera? The correspondence between some of the synonyms of frankincense and the words used in different languages for the slaying or burning of sacrificial victims is very interesting. But I have only to observe, in connexion with this, two things that have always struck me in Bombay, viz., that the burning of the dead, in the form of the pyre, and the manner in which the body is burned, very forcibly suggests that the practice is derived from human bloody sacrifice, and that the use of sweet odours, in religious rites generally, originated in sanitary precautions. They do not merely mask bad smells, but correct them; and they wonderfully refresh the spirits from the depression which they fall into in crowded places like churches. They also drive out vermin. Nothing so quickly clears your bed in Bombay of mosquitos as burning a little olibanum or myrrh in it. The Protestant churches there are infested by mosquitos as by a plague; and I was often asked how to get rid of them, and always answered that the ugly and irreverent punkahs were of no use, and that the only remedy was the immemorial and most beautiful rite of burning frankincense. A pot of holy basil is placed before every Hindu home; and the mother of the house may be seen every morning, after having ground the corn for the day, and performed her simple toilet, walking round and round the four-horned altar, on which the pot of holy basil is set, invoking the blessings of heaven on the father of her children and on them.

DESCRIPTIO.

GENUS BOSWELLIA.

Roxburgh, Pl. Corom. iii. p. 4, t. 207; Benth. & Hook. Gen. Plant. i. p. 322. [*Libanus*, Colebrooke in As. Res. ix. p. 377, t. 5. f. 1. *Plösslia*, Endlicher, Nov. Stirp. Dec. 47; Iconog. t. 119, 120.]

Calyx parvus, plerumque 5- interdum 6- non sæpe 7-, rarissime 4- v. 10-dentatus, persistens. *Petala* tot quot calycis divisiones, basi angustata, patentissima, imbricata. *Discus* annularis, crenatus, campanulatus v. explanatus. *Stamina* sepalis duplo numerosiora, plerumque 10, brevia, basi disci inserta, alterna interdum breviora. *Ovarium* 3-, creberrime 4-, interdum 5-loculare; stylus brevis, stigmate 3-, 4- v. 5?-lobo; ovula in loculis 2, pendula. *Drupa* 3-, 4-, 5-gona, 3-, 4-, 5-valvis, valvis a pyrenis secedentibus; pyrenis osseis, 1-spermis, ab axi persistente solutis. *Semina* compressa, pendula, margine membranaceo, testa membranacea; cotyledones contortuplicatæ, multifidæ; radicula supera.

Arbores cortice sæpe papyraceo, thuriferæ. *Folia* decidua, apices versus ramulorum conferta, patentia, alterna, imparipinnata, foliolis oppositis, serratis v. integerrimis. *Racemi* v. *paniculæ* axillares v. terminales, simplices v. compositi. *Flores* albi, v. rubicundi, v. virides, v. vix luteoli, v. cerei.

Species 5, Nubiæ, Abyssiniæ, Berberæ, Arabiæ, Indiæque incolæ.

Sectio I. *Plantæ plus minus pubescentes, disco campanulato.*

Subsectio *a*. *Racemi simplices.*

1. BOSWELLIA CARTERII, Birdwood, n. sp. (Tab. I. and II. fig. 1.)

DIAGNOSIS.—*Ramuli* terminales, pubescentes vel tomentosi; *foliola* 7–10-juga, ovato-oblonga, undulata vel crenato-undulata, late rotundata vel basi truncata. *Racemi* simplices, fasciculati, foliis breviores. *Fructus* ¼ ad ½ poll. longus, obovato-obtusus.

Carter, Journ. Bombay Br. Roy. As. Soc. ii. 1847. *B. thurifera*, Colebrooke?

Habitat in montibus Soumali, "Mohr Madow" incolarum, *Playfair*; et in montibus Hadramaut in Arabia, "Maghrayt d'sheehaz" apud Maharas, *Carter*.

Descriptio.—Arbor parva; petiolus pubescens; foliola aut crenato-undulata et omnsa pubescentia, aut obscure serratula vel integerrima et glabra supra, et velutina interdumque pallidiora infra, longa circa 1½ poll. et lata circa ¾ poll., foliolo terminali sæpe grandiore. Corolla patens, petala alba vel cerea, discus roseus, ovarium disco carnoso reseo arcte adpresso semiimmersum.

Remarks.—The above description is taken from the specimens of leaves collected for Playfair from the "Mohr Meadow" of the Soumalis, and the young plants growing in the Victoria Gardens from cuttings of the same, and the flowers gathered by Playfair in June 1867 from the young plants growing in the Aden Garden from cuttings of the same, and sent by him to the Herbarium at Kew.

The tree found by Carter growing in the limestone formation about Merbat and Ras Fartak, the "Maghrayt d'sheehaz" of the Maharas, is probably a smaller tree in habit. The plant which he brought from the Hadramaut to Bombay, is to this day growing there in the Victoria Gardens; and whilst it is almost impossible to distinguish its leaves from those of the "Mohr Madow" of

[1] "First Footsteps in East Africa," London, 1856. In the preface to this work, Burton says that an expedition into the Soumali land was first proposed by Sir Charles Malcolm in 1849, and that the command of it was offered to and accepted by Carter in 1851, but that after Sir Charles Malcolm's death the project was given up; and that when Burton proposed his expedition in 1854 Stocks was appointed to it as botanist, but died before it started. The *savans* who went with the "Abyssinian Army" had no time for work, so swift was Lord Napier's march to Magdala and back.

the Soumalis, except from their being smaller and more crumpled, it is not more than half the height of the plants of "Mohr Madow" propagated in the Victoria Gardens from the cuttings sent to Bombay by Playfair, and which are already about 9 feet high. In "Maghrayt d'sheehaz," the leaves have sometimes five pairs only of leaflets.

Playfair's dried specimens of leaves of "Mohr Madow," sent to Kew and Bombay, include two very distinct varieties,—one crenate, undulate, and pubescent on both sides; the other undulate and obscurely serrulate or almost entire, and velvety and paler below and glabrous above. The plants of "Mohr Madow" in the Victoria Gardens and in Playfair's Garden at Aden have as yet shown only the former variety of leaf, and the plant "Maghrayt d'sheehaz" in the Victoria Gardens never has shown any other; but from the analogy of *B. thurifera* and *B. Bhau-Dajiana* ("Mohr Add") I am led to believe that the two kinds of leaves included in Playfair's dried specimens of "Mohr Madow" are only varieties of leaf of one species, and not leaves of two distinct species of *Boswellia*. I have therefore included the description of both leaves in my description of *B. Carterii*, but *not without uncertainty*.

[(*Added* 1874.) CARTER's plant ("*Maghrayt d'sheehaz*") was not before me when I was preparing this paper (see pp. 138, 139), and my description of *B. Carterii* was, and is, the description of "*Mohr Madow.*" A flowering specimen of "*Maghrayt d'sheehaz*" was subsequently received from Mr. Gray, B.M., the present Director of the Victoria Gardens, Bombay, when it was found that in it *the stamens are inserted into the side of the disk*, but it was too late to add the character to the description of the species *Boswellia Carterii* and of the genus. This character therefore distinguishes the Arabian (*Maghrayt d'sheehaz*) from the Soumali (*Mohr Madow*) variety of *B. Carterii.*]

2. BOSWELLIA BHAU-DAJIANA, Birdwood, n. sp. (Tab. III.)

DIAGNOSIS.—*Ramuli* terminales, modice puberuli vel glabri; *foliola* 7–10-juga, oblongo-lanceolata vel oblonga, late rotundata vel basi truncata, aut serrata et pubescentia, aut obscure serrulata et infra incana, supra glabra. *Inflorescentia* in simplicifasciculatis racemis. *Fructus* immaturus oblongo-contractus basi.
Habitat in montibus Soumali, "Mohr Add" incolarum, *Playfair.*

Descriptio.—Arbor parva; folia aut pubescentia, foliolis serratis, aut supra glabrà et infra incana foliolis obscure serratulis vel integerrimis, longa 1 poll. ad 1½ et 2 poll. et lata ⅛–½ poll.; 5 vel 6 racemi in quoque fasciculo, foliis paulo breviores; flores numerosissimi, corolla vix patentissima, petala alba v. viridula; discus valde pubescens, virescens, ovarium disco concavo haud adpresso semiimmersum; drupa subclavata.

Remarks.—Playfair's dried specimens of leaves of "Mohr Add," sent to Bombay, include two distinct varieties, the leaflets in one being serrulate and downy, in the other obscurely and remotely serrulate or entire, and hoary beneath, and glabrous above. The latter is the only variety of leaf sent by him to Kew. But in one of his dried specimens of "Mohr Add" sent to Bombay, both these varieties of leaf are seen in the same specimen. The plants in the Victoria Gardens all showed the pubescent and serrate variety of leaf until last year, when one threw out both the varieties of leaf described, and a third variety also of intermediate character. The plant from which I gathered the flower represented in my plate of *B. Bhau-Dajiana* showed at that time only the serrate and pubescent variety of leaf; but I have figured beside it two other varieties of leaves which I subsequently found *either on this same plant, or other plants of the same species growing near it. It would be more satisfactory if I were less uncertain on this point*, especially as my ground for including the two kinds of "Mohr-Madow" leaves sent to me by Playfair under the species *B. Carterii* is the all but absolute certainty that analogous varieties of leaf are produced by *B. Bhau-Dajiana* or "Mohr Add." It must be distinctly borne in mind, therefore, that with the leaves no flowers have ever been seen of the variety, either of "Mohr Madow" or "Mohr Add," in which the leaflets are more or less entire, and hoary below, and glabrous above, and that it may be an error to include them as varieties respectively of *B. Carterii* and *B. Bhau-Dajiana.* But analogous variations of leaf are conclusively authenticated in the case of *B. thurifera, infrà.*

The serrate downy variety of leaflet in *B. Bhau-Dajiana* is very like the young leaflets of *B. Carterii;* but the leaflets in the latter, young and old, are always more or less undulate, whereas in the former they are always quite flat. In the former, also, the leaflets are narrow and more lanceolate, in the latter broad and more oval; and in the mass of foliage these differences make a distinction.

B. Carterii [*i.e.* "Mohr Madow"] and *B. Bhau-Dajiana* ["Mohr Add"] yield the bulk of the olibanum exported from the Soumali country under the name of "Laban-Sheheri," "possibly," writes Playfair, "because it is principally taken to Sheherr and Makalla markets." Vaughan[1] says "Luban Shaharee" is the term given to Arabian frankincense, which is also yielded by *B. Carterii* [*i.e.* "Maghrayt d'sheehaz"]. Both the glabrous-above hoary-beneath variety (?) of *B. Carterii* [*i.e.* "Mohr Madow"] and of *B. Bhau-Dajiana* ["Mohr Add"] yield, according to Playfair, the "Luban Bedowi" of the Soumalis, which is also mentioned by Cruttenden, but nowhere by Vaughan.

Subsectio β. *Racemi compositi.*

3. BOSWELLIA PAPYRIFERA, Richard, Tent. Flor. Abyss., Voyage en Abyssinie par Lefebre, &c., iv. p. 140, tab. 33.

DIAGNOSIS.—*Foliola* 4–10-juga, pubescentia. *Inflorescentia* racemosa, racemis paniculatis.
Amyris papyrifera, Delile, Cent. Plant. d'Afrique; Cailliaud, Voyage à Méroé, ii. p. 293.

---

[1] Pharm. Journ. xii. *sup. cit.*

Plösslea floribunda, Endlicher, Nov. Stirp. Mus. Vind. Dec. no. 47, tab. 119, 120.
Bruce, Travels, vii. p. 334, tab. 48, "Angouah."
Habitat in Senaar et Soudan, et Gondar, in convallibus fluviorum Taccaze et Mareb, "Angouah" incolarum, *Bruce*; "Kafal Galgalaam" et "Loban-adam-coulan," *Delile*; "Makker," *Dillon* et *Schimper* apud *Richard*.

Descriptio.—Arbor magna; foliola lanceolata, acuta, dentata, longa 1 poll. ad 2 et 3 poll.; calyx longe dentatus; corolla reflexa; petala longa, reflexa; pistillus basi contractus; drupa clavata.

Remarks.—Of the tree named "Angouah" in Bruce's "Travels," he says, "The Abyssinians "believe it to be that which bears the true frankincense; in reality it produces a gum much "resembling it." Richard says of *B. papyrifera*, "It distils a resin which, when burnt, gives the "odour of the incense of India, which, indeed, as every one says, is produced by another species "of the genus, named *B. serrata*, Roxb."

4. Boswellia thurifera, Colebrooke, Asiatic Researches, ix. p. 377, t. 5.

Diagnosis.—*Foliola* 8–14-juga, dentata, serrata, vel plus minus obscure et remote serrulata, pubescentia v. glabra. *Inflorescentia* racemosa, racemis compactis, plerumque dense compositis, foliis multo brevioribus. *Fructus* late ellipticus, triquetrus, basi vel apice non contractus.
B. serrata et *B. glabra*, Roxburgh, Flor. Ind. ii. p. 383, 384.
B. glabra, Roxburgh, Cor. ii. t. 207.

Var. α. Foliola ovato-oblonga, truncata vel late rotundata basi, serrata, infra pubescentia.
Var. β. Foliola lineari-lanceolata, oblique contracta basi, nonnunquam obscurissime vel remote serrulata, propemodum glabra.

Habitat in montibus Indiæ tropicæ, "Salai," "Saleh" incolarum; Oude et Rohilcund, *Royle*; Behar, *Hooker*; Concan, *Stocks*; Kattyawar, Khandeish, *H. M. Birdwood.*

Descriptio.—Arbor parva; petala patentia, infra pubescentia, valde rosacea, v. albida; stamina alterna breviora; discus ruber, v. inconspicuus.

Remarks.—The natives of India recognize the two varieties of this plant, of which Roxburgh made two species, and distinguish between their gum-resins. What I have seen of it has been either stalactiform, like the runnings of a wax candle, or in small tears, and always so soft that, when kept in a bottle, in a short time it has run into an oleo-resinous mass, with the smell of frankincense, but more turpentiny. I often and perseveringly tried to get regular tears of olibanum from this plant, but never succeeded in getting anything else than soft, oleo-resinous "runnings" from it, which, even after months' exposure on the trunk, still remained quite soft.

In Khandeish the olibanum produced by this plant is sold under the name of "Dup-Salai" [*i.e.* incense of Salai] in the village bazaars. But I have never myself seen it exposed for sale. I have had samples sent to me from all parts of India, and all soft and always melting at last into a mass.

The most remarkable character of this plant is the way in which its leaves and *flowers* vary. Sometimes the inflorescence is in loose cymes, sometimes gathered into a knotty head or capitulum. In a plant brought by my brother from Kattyawar, the petals were reduced to mere scales, the disk very fleshy, red, and prominent, and the stamens tending to abortion.

Sectio II. *Frutex glaber, glaucus, disco explanato.*

5. Boswellia Frereana, Birdwood, n. sp. (Tab. II., fig. 2, and IV.)

Diagnosis.—*Folia* omnia glabra et glauca; foliola 3–6-juga, late ovata, elliptica, valde obtusa utrinque, undulata. *Inflorescentia* in racemis compositis, foliis multo longior. *Fructus* ovato-oblongus.
Habitat in montibus calcareis Soumaliensis regionis prope "Bunder Murayha," *Cruttenden, Kempthorne*, et *Vaughan*: "Yegaar" Soumaliensium, *Playfair.*

Descriptio.—Arbor parva; folia et flores omnia citrum et terebinthum grate redolentia; corolla valde et rigide patens, stellata; petala basi contracta, et acuta, smaragdina; discus explanatus; stamina circa marginem inserta, rigide erecta in floribus recentioribus, et super disco supine jacentia (velut radii rotæ) in deflorescentibus; antheræ rubræ; pistillus omnino liber et super discum conspicue elevatus, velut metula.

Remarks.—This is a strongly marked species, the source of the "Louban Maitee" of the Soumalis. It is unmistakably this plant which Kempthorne describes. It is evidently also Cruttenden's "Luban Meyti."

Kempthorne says, "The tree is one of the most extraordinary plants I ever saw, quite a *lusus naturæ* of the vegetable world, for the trees actually grow out of the sides of the almost polished rocks. . . . . The trees were about 40 feet high, the stem was about 2 feet in circumference, rising straight up, with a bend outwards of 6 or 7 inches. They are attached most firmly to the rocks by a thick oval mass of substance, about a foot or so in diameter, something resembling a mixture of lime and mortar. Branches spring out rather scantily at the top, and extend a few feet down the stem; the leaves are 5 inches or so long, and 1½ broad, narrowing and rounding towards the point, but not serrated at the edges; the upper surface is of a rich dark shining green, while the lower is of a lighter hue; they are thin and smooth, and crimped like that beautiful species of seaweed so often found on the coast of England. The tree has four layers of bark, the outer being coarse and loose, like that of the beech, while the next two are as it were glued to the trunk and delicately fine, resembling oiled paper or gold-beater's skin, and of a bright amber-colour; this bark is perfectly transparent, and can be stripped off easily in large sheets; the natives use it for writing on; the inner bark of all is an inch or so in thickness, adhering closely to the stem; it is tough, not unlike leather, and striped red and white, and yields a strong aromatic perfume. The timber is white, soft, porous, and of little use, except as firewood. A deep incision into the bark causes the odoriferous gum to exude in large quantities, which is of a milky white, and of the consistency of honey; but it soon hardens by exposure to the atmosphere. It is a remarkable fact that not a single frankincense-tree did I perceive growing upon any other rocks than those of almost pure limestone . . . . It is my opinion that the trees on the Adel coast are different from the species as described so ably by Dr. Carter" (Trans. Bombay Geo. Society, xiii. 1857). This description of Kempthorne's well shows how we should understand the descriptions of old writers. The present Governor-General of Mozambique informed me, quite in the style of Sir John Maundeville, "that copal grew in the earth just like potatoes."

Vaughan writes:—"The hill-ranges on the eastern coast of Africa are composed entirely of white limestone, in some parts so compact as to resemble alabaster. This appears to be the soil most genial to the tree, and in no instance did I find it growing in sand or loam. The tree is first met with at a few miles inland from the coast, and at an altitude of about 300 feet above the level of the sea. Its appearance is strikingly singular, seeming at first sight to be destitute of roots and clinging to the hard uncreviced rock by masses of a rhomboidal and fantastically shaped wood with the most obstinate adherence. The stem is nearly at right angles with this substance, ascending almost invariably in an upright direction, and attaining the height of from 12–15 feet. At the base the circumference is equal to that of a man's thigh, gradually tapering towards the top, where it shoots off its branches and leaves" (Pharm. Journ. and Trans. xii. 1853).

All the frankincense trees have a general resemblance to the Mountain Ash, when putting forth its spring leaves, and especially *Yegaar*. As I saw this plant in Playfair's garden at Aden in September last, in young leaf and covered with bloom, I was much struck by its elegant singularity. The long racemes of green star-like flowers, tipped with the red anthers of the stamens (like aigrettes of little stars of emerald set with minute rubies), droop gracefully over the clusters of glossy glaucous leaves; and every part of the plant (bark, leaves, and flowers) gives out the most refreshing lemon-like fragrance, resembling that of the blue flowered *Salvia patens*.

It gave me the most lively pleasure to find the flowers which I had been so long seeking of so charming an originality, and in such a romantic spot; for Playfair's garden is a mere angle[1] in the shadow of two pumice rocks, which tower hundreds of feet above it, guarding in their shadows[2] the store of water for the garrison and town; and green leaves, shade, and water make Paradise in the east. My pleasure was made greater because of the presence of two priests of the Society of Jesus who walked in the shadow, amongst the tanks and green boughs, from the memories they recalled of old learned books, more grateful than the fragrance of the lemon-scented *Yegaar* of the Edens of Adel.

---

[1] " . . . . et quod
Angulus iste feret piper et thus ocius uva."
Hor. Epist. lib. I., xiv. 22–23.

[2] Herbert, Travels, Lond. 1677, describes Aden as "a town low in its situation and sulphureously shaded by the high and barren mountain Cabobarra, whose brazen front so scorches this miserable place."

PLATE I.

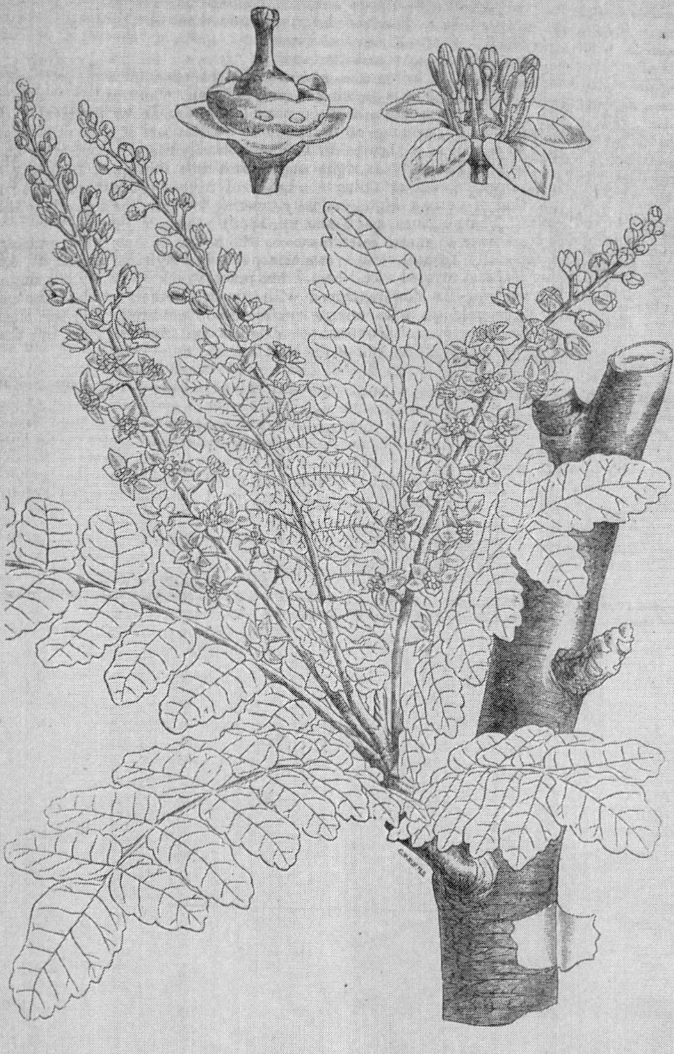

BOSWELLIA CARTERII. *Birdw.*

var.—Maghrayt d'Sheehaz of the Maharas.

PLATE II.

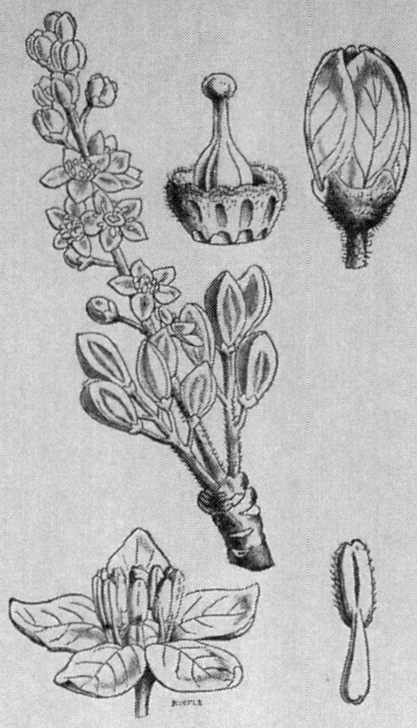

Fig. 1. Boswellia Carterii. *Birdw.*
var.—Mohr Madow of the Soumalis.

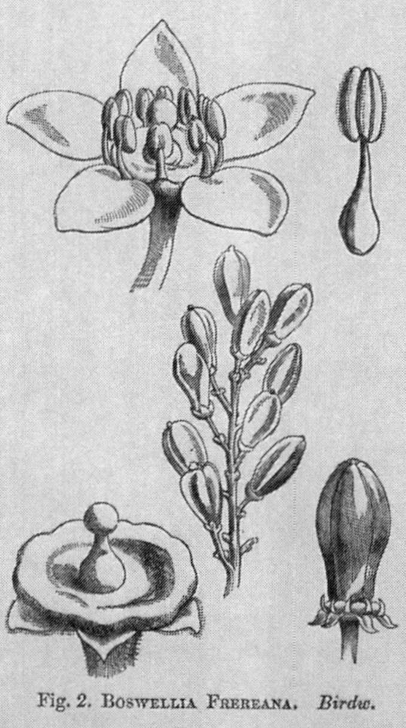

Fig. 2. Boswellia Frereana. *Birdw.*

PLATE III.

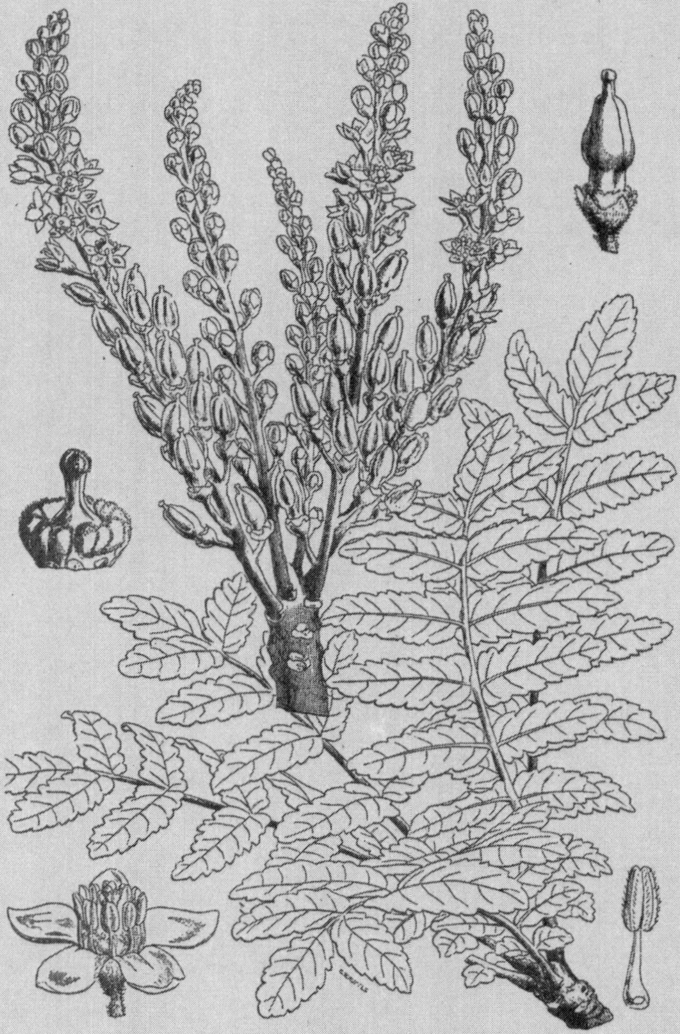

BOSWELLIA BHAU-DAJIANA. *Birdw.*

Mohr-Add of the Soumalis.

PLATE IV.

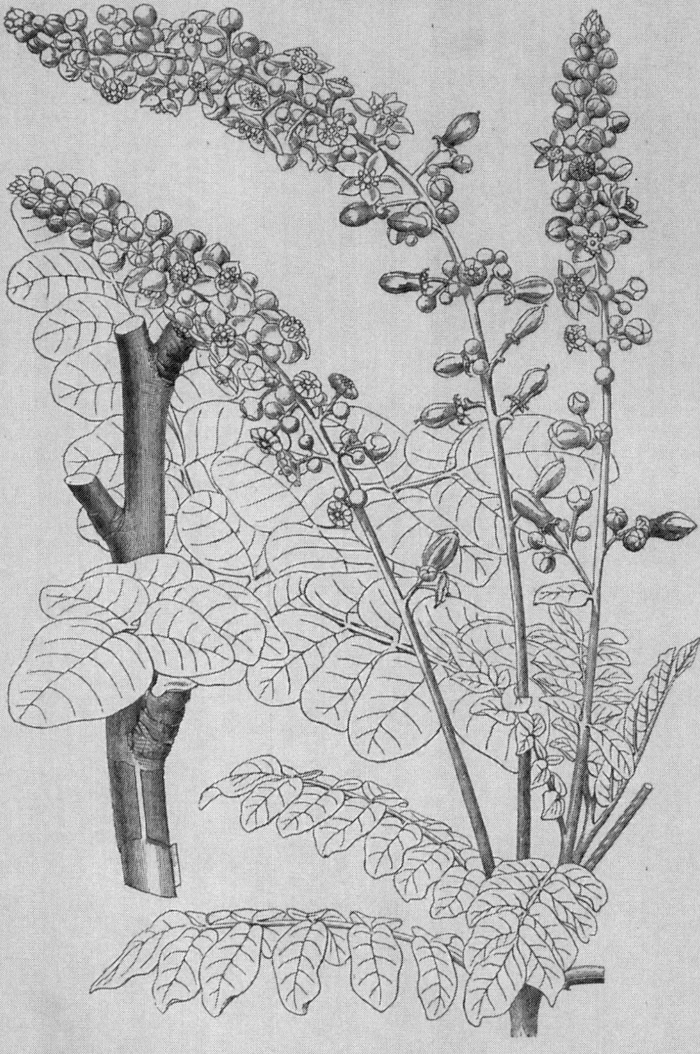

BOSWELLIA FREREANA. *Birdw.*

Yegaar of the Soumalis.